Oxyfuel Gas Welding

by
Kevin E. Bowditch • Mark A. Bowditch • Ronald J. Baird

Publisher
The Goodheart-Willcox Company, Inc.
Tinley Park, Illinois

Using This Write-in Text

Oxyfuel Gas Welding presents the fundamentals of this skilled trade in an easy-to-understand manner. Each unit includes problems for you, the student welder, to solve.

Oxyfuel gas welding requires a great deal of practice to produce satisfactory weldments. To become a good welder, you will have to spend many hours running beads and making the joints presented in this write-in text. Read the text matter carefully, study and complete the questions, and perform the activities as required. This will aid you in developing skills and techniques that will enable you to enter the welding industry and to prepare yourself for many career opportunities.

Each practice piece should be evaluated, and every effort should be made to correct any problem encountered. Where possible, test each weld for integrity, penetration, and appearance. Your instructor will help you do this. To conserve material, it is recommended that both sides of the practice piece be used whenever practical. Identify your work by stamping your name or initials on each practice piece.

The questions in the Check Your Progress section at the end of each unit will help you determine how well you understand the information provided. Additional up-to-date welding know-how can be acquired by carrying out the suggested activities.

The Goodheart-Willcox Company, Inc. Brand Disclaimer: Brand names, company names, and illustrations for products and services included in this text are provided for educational purposes only, and do not represent or imply endorsement or recommendation by the author or the publisher.

Goodheart-Willcox Publisher Safety Notice: The reader is expressly advised to carefully read, understand, and apply all safety precautions and warnings described in this book or that might also be indicated in undertaking the activities and exercises described herein to minimize risk of personal injury or injury to others. Common sense and good judgment should also be exercised and applied to help avoid all potential hazards. The reader should always refer to the appropriate manufacturer's technical information, directions, and recommendations; then proceed with care to follow specific equipment operating instructions. The reader should understand these notices and cautions are not exhaustive.

The publisher makes no warranty or representation whatsoever, either expressed or implied, including but not limited to equipment, procedures, and applications described or referred to herein, their quality, performance, merchantability, or fitness for a particular purpose. The publisher assumes no responsibility for any changes, errors, or omissions in this book. The publisher specifically disclaims any liability whatsoever, including any direct, indirect, incidental, consequential, special, or exemplary damages resulting, in whole or in part, from the reader's use or reliance upon the information, instructions, procedures, warnings, cautions, applications or other matter contained in this book. The publisher assumes no responsibility for the activities of the reader.

Library of Congress Cataloging-in-Publication Data
Bowditch, Kevin E.
Oxyfuel gas welding / by Kevin E. Bowditch, Mark A. Bowditch, Ronald J. Baird.
   p. cm.
  Includes index.
  ISBN 1-59070-300-6
  1. Oxyacetylene welding and cutting. I. Bowditch, Mark A.
II. Baird, Ronald J.   III. Title.
TS228.B59 2004   2003055362
671.5'22—dc22   CIP

Contents

Brazing Techniques

Soldering Techniques

Technical Information

Company Acknowledgments

Airco Welding Products

Airco Welding Products, Div. Air Reduction Co.

ALCOA

American Torch Tip Co.

American Welding Society

Handy & Harman/Lucas-Milhaupt, Inc.

Linde Co. Div. of Union Carbide Corp.

Nederman, Inc.

Smith Welding Equipment Co.

Uniweld Products, Inc.

Victor Equipment Co.

Unit I

INTRODUCTION TO OXYACETYLENE WELDING

Welding, Cutting, and Joining Processes

Welding and cutting are processes used in industry to join and cut metals. Some processes are manual and require simple equipment. Others are automatic and make use of complicated equipment and processes like lasers and robots.

Many different welding and cutting processes are used to meet the needs of modern industries. See **Figure 1-1.** Arc welding and resistance welding are the processes most often used. Two other important processes are oxyfuel gas welding and cutting.

Different fuel gases can be combined with oxygen to create the heat necessary for welding, cutting, and joining.

Master Chart of Welding, Joining, and Allied Processes

Atomic hydrogen welding........ AHW
Bare metal arc welding........... BMAW
Carbon arc welding................. CAW
-gas................... CAW-G
-shielded..................... CAW-S
-twin........................ CAW-T
Electrogas welding.................. EGW
Flux cored arc welding........... FCAW
-gas shielded................. FCAW-G
-self-shielded.................. FCAW-S

Gas metal arc welding............... GMAW
-pulsed arc........................... GMAW-P
-short circuiting arc............ GMAW-S
Gas tungsten arc welding........... GTAW
-pulsed arc........................... GTAW-P
Magnetically impelled
arc welding........................ MIAW
Plasma arc welding.................... PAW
Shielded metal arc welding......... SMAW
Arc stud welding........................ SW
Submerged arc welding.............. SAW
-series................................ SAW-S

Coextrusion welding............CEW
Cold welding.........................CW
Diffusion welding..................DFW
Explosion welding.................EXW
Forge welding.......................FOW
Friction welding....................FRW
-direct drive.....................FRW-DD
-friction stir....................FSW
-inertia friction...............FRW-I
Hot pressure welding...........HPW
-isostatic........................HIPW
Roll welding.........................ROW
Ultrasonic welding...............USW

Block brazing........................... BB
Diffusion brazing...................... DFB
Dip brazing.............................. DB
Exothermic brazing.................. EXB
Furnace brazing....................... FB
Induction brazing...................... IB
Infrared brazing....................... IRB
Resistance brazing................... RB
Torch brazing.......................... TB
Twin carbon arc brazing........... TCAB

Dip soldering........................... DS
Furnace soldering.................... FS
Induction soldering.................. IS
Infrared soldering....................IRS
Iron soldering..........................INS
Resistance soldering.............RS
Torch soldering........................ TS
Ultrasonic soldering................. USS
Pressure gas soldering......... WS

Adhesive bonding............ AB
Braze welding.................. BW
-arc................................ ABW
-carbon arc.................... CABW
-electron beam........... EBBW
-expthermic.................. EXBW
-flow brazing............. FLB
-flow welding............. FLOW
-laser beam.............. LBBW
Electron beam welding..... EBW
-high vacuum.............. EBW-HV
-medium vacuum....... EBW-MV
-nonvacuum............... EBW-NV
Electroslag welding........... ESW
-consumable guide.... ESW-CG
Flow welding.................... FLOW
Induction welding............. IW
Laser beam welding......... LBW
Percussion welding......... PEW
Thermite welding............. TW

Flash welding....................... FW
Pressure-controlled
resistance welding......... PC-RW
Projection welding............... PW
Resistance seam welding..... RSEW
-high frequency............. RSEW-HF
-induction..................... RSEW-I
-mash seam.................. RSEW-MS
Resistance spot welding....... RSW
Upset welding...................... UW
-high frequency............. UW-HF
-induction..................... UW-I

Air acetylene welding......AAW
Oxyacetylene welding.....OAW
Oxyhydrogen welding......OHW
Pressure gas welding..... PGW

Arc spraying...........................ASP
Flame spraying......................FLSP
-wire..............................FLSP-W
High velocity oxyfuel
spraying......................HVOF
Plasma spraying.....................PSP
Vacuum plasma spraying.....VPSP

Carbon arc cutting........... CAC
-air carbon arc cutting... CAC-A
Gas metal arc cutting........ GMAC
Gas tungsten arc cutting... GTAC
Plasma arc cutting........... PAC
Shielded metal arc cutting. SMAC

Flux cutting............................ OC-F
Metal powder cutting............... OC-P
Oxyfuel gas cutting.................. OFC
-oxyacetylene cutting..... OFC-A
-oxyhydrogen cutting...... OFC-H
-oxynatural gas cutting... OFC-N
-oxypropane cutting....... OFC-P
Oxygen arc cutting................. OAC
Oxygen gouging...................... OG
Oxygen lance cutting............ OLC

Electron beam cutting........EBC
Laser beam cutting............LBC
-air.............................. LBC-A
-evaporative................. LBC-EV
-inert gas.................... LBC-IG
-oxygen......................LBC-O

Arc welding (AW)

Solid-state welding (SSW)

Brazing (B)

Soldering (S)

Welding and joining processes

Other welding and joining

Resistance welding (RW)

Oxyfuel gas welding (OFW)

Thermal spraying (THSP)

Allied processes

Oxygen cutting (OC)

Thermal cutting (TC)

Arc cutting (AC)

High energy beam cutting

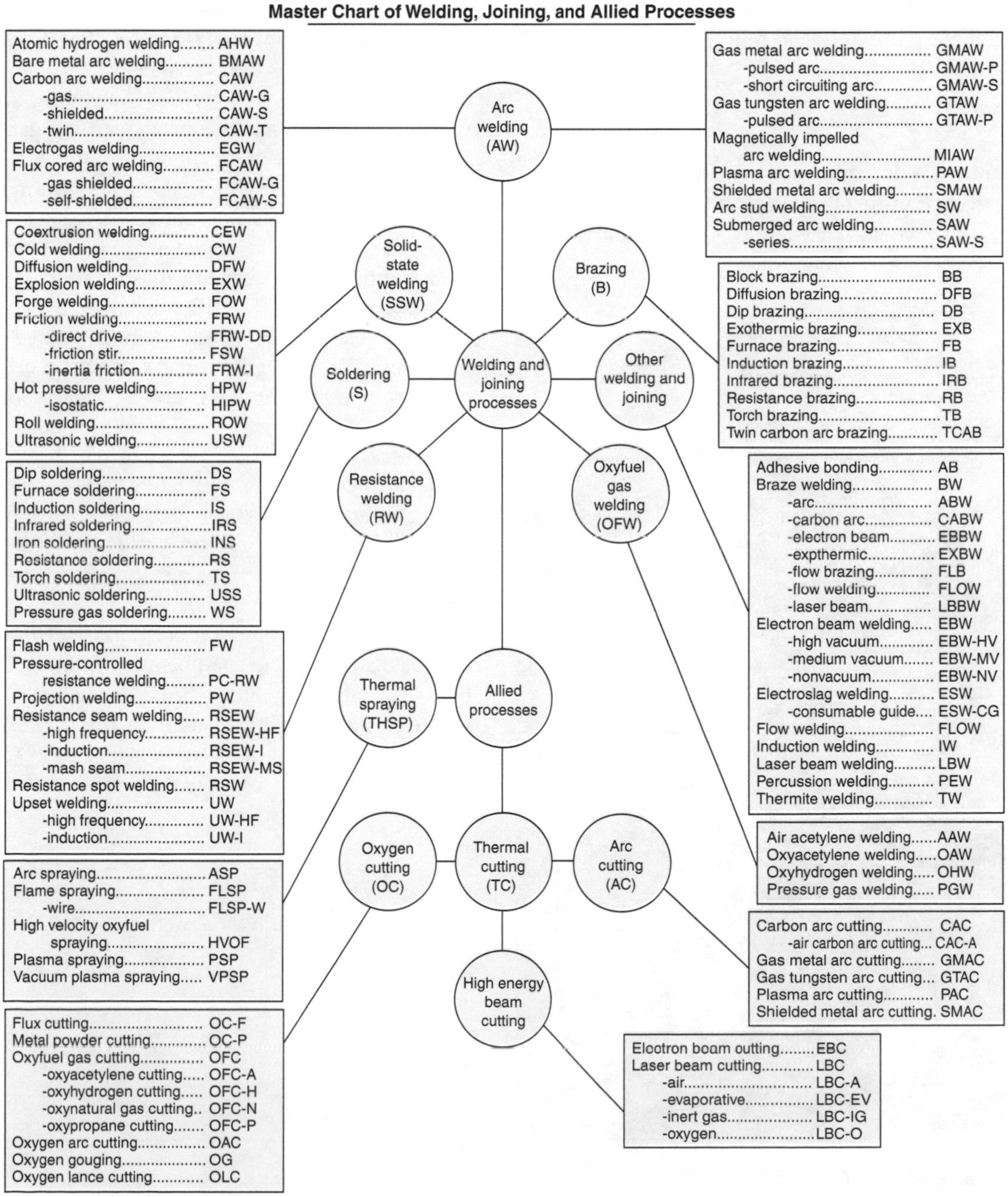

Figure 1-1. The many welding processes used in industry. (AWS) (Adapted from AWS A3.0:2001)

Acetylene is one such fuel gas. When properly burned, acetylene and oxygen produce a hotter flame than any other fuel gas/oxygen combination. Therefore, acetylene and oxygen are most often used. The combination is called oxyacetylene.

This textbook covers oxyacetylene welding and the related processes of oxyacetylene cutting, torch brazing, and torch soldering. If a fuel gas other than acetylene is used, the process is referred to as oxyfuel gas welding or cutting.

An oxyacetylene torch with a cutting attachment is a handy tool, **Figure 1-2.** A flame temperature of 5589°F (3087°C) is obtained with a neutral oxyacetylene flame. This provides enough heat for welding and cutting steel, the two major uses of the oxyacetylene welding outfit. However, it also provides a source of heat for brazing, soldering, and heat treating metals.

Since this text covers welding, flame cutting, brazing, and soldering of metals, it is important to understand the differences between these processes. The following definitions will briefly explain those differences.

Process Definitions

The American Welding Society defines a weld as: "A localized coalescence of metals or nonmetals produced either by heating the materials to the welding temperature, with or without the application of pressure, or by the application of pressure alone and with or without the use of filler metal."

Oxyfuel gas welding refers to the use of oxygen with fuel gases such as acetylene, hydrogen, natural gas, butane, or propane. Oxygen and hydrogen is the most common combination used in industry.

Oxyacetylene welding and *arc welding* both produce a weld by heating the base metals until they become liquid. Another term for liquid is *molten.* The molten metals flow together in the weld pool. When the liquid metal solidifies (becomes solid), a weld is produced. Heating two or more metals or nonmetals until they become liquid, then allowing them to flow together and solidify, is called *fusing.* Thus, oxyfuel gas welding and arc welding are called fusion welding processes.

Fusion welding involves joining metals (or nonmetals) by heating them to their melting points. When a suitable temperature is reached, the materials melt and fuse (stick) together into one piece. Welding is done with or without the use of filler metal. Filler metal, when used, has almost the same material and melting point as the base metals to be joined.

Flame cutting is a process that uses high-temperature oxidation (combining the metal with oxygen) to cut metal. Heat is produced by an oxyfuel gas flame. A stream of oxygen burns or oxidizes away the base metal to form a cut.

Brazing includes a group of processes in which metal pieces to be joined are heated above 840°F (450°C) but below their melting points. A nonferrous braze filler metal is used to join the base metals. The filler metal has a melting point above 840°F (450°C) and below the melting point of the base metals. The braze filler metal fills small areas between mating parts by capillary action. *Capillary action* occurs when a liquid is drawn into a small gap between mating parts.

Braze welding is a process in which metal pieces to be joined are heated above 840°F (450°C) but below their melting points. A braze filler metal with a melting point above 840°F (450°C) is used to fill a groove weld or to create a fillet or other weld bead. The braze filler metal does not fill the joint using capillary action.

Soldering is a joining process in which the metal pieces to be joined are heated to a temperature below 840°F (450°C) and below the melting point of the base metal. A filler metal called *solder* is used to fill the small spaces between mating parts. Solder has a melting point below 840°F (450°C) and below the base metal. Molten solder is drawn into the joint by capillary action.

Check Your Progress

Write your answers in the spaces provided.

1. Using the chart in Figure 1-1, identify and list the oxyfuel gas welding processes.

    a. _____

    b. _____

    c. _____

    d. _____

Figure 1-2. An oxyacetylene cutting torch is used to cut pipe. (Victor Equipment Co.)

2. In your own words, briefly describe how brazing differs from soldering.

   _____

   _____

   _____

   _____

3. In your own words, briefly describe how brazing differs from braze welding.

   _____

   _____

   _____

   _____

4. What are the two major uses of oxyacetylene equipment?

   a. _____

   b. _____

5. Identify machines, tools, or products around your shop that have been fabricated by welding. Explain why welding was chosen over some other fastening device. __

   _____

   _____

   _____

   _____

   _____

Instructor's Initials _____ Date _____

Things to Do

1. Go to your local library and familiarize yourself with the many books and magazines related to the field of welding.

2. Use the Internet (available at most libraries) to obtain online information about the field of welding. Begin by choosing a search engine and keying the words welding or American Welding Society.

3. Become a member of the American Welding Society, and/or subscribe to various welding magazines to expand your knowledge of the field.

Unit 2

WELDING SAFETY

Safety First

Oxyfuel gas welding equipment can be very dangerous. Welders must always be aware of potential hazards to themselves and others. Equipment, tools, protective clothing, and safety devices must be used properly. Precautions have been put in place to keep workers and the work environment safe. By routinely observing basic safety precautions, the welder will develop an attitude of "safety first." This attitude will encourage others to do the same.

Personal Protection

The most important personal protective device in the welding area is eye protection. When metal cutting, machining, grinding, and welding, always wear approved safety glasses or goggles to protect your eyes from flying particles.

Goggles with colored filter lenses block out harmful ultraviolet and infrared rays, **Figure 2-1.** They come in a range of shades, numbered 1 to 14. The higher the number, the darker the lens, **Figure 2-2.** The higher numbered lenses are required when arc welding with high currents. **Figure 2-3** shows the filter and clear cover lenses of a pair of goggles.

Never look directly at an oxyfuel gas flame or arc if your eyes are unprotected! Wear goggles with the correct filter lens when lighting, welding, or cutting with an oxyfuel torch. Wear a welding helmet with the correct filter lens when arc welding or cutting.

Do not wear contact lenses in a welding area! Contact lenses are designed to rest on the fluid of the eyes. During welding, it is possible for smoke, fumes, and dust particles to lodge between the lens and eye. Furthermore, a person wearing contact lenses should never be exposed to an arc flash without proper filter lenses.

Process	Thickness of Material	Shade Number
Oxyfuel Gas Welding		
Light	Under 1/8″ (3.2 mm)	4 or 5
Medium	1/8″ to 1/2″ (3.2 mm to 12.7 mm)	5 or 6
Heavy	Over 1/2″ (12.7 mm)	6 or 8
Oxyfuel Gas Cutting		
Light	Under 1″ (25.4 mm)	3 or 4
Medium	1″ to 6″ (25.4 mm to 150 mm)	4 or 5
Heavy	Over 6″ (150 mm)	5 or 6
Torch Brazing		3 or 4
Torch Soldering		2
Shielded Metal Arc Welding		8 to 14
Gas Tungsten Arc Welding		8 to 14
Gas Metal Arc Welding		8 to 14

Figure 2-2. Filter lens shade numbers for the most common welding applications. A filter lens protects the eyes from ultraviolet and infrared radiation. Higher numbered filter lenses provide more protection and are required when arc welding with higher currents.

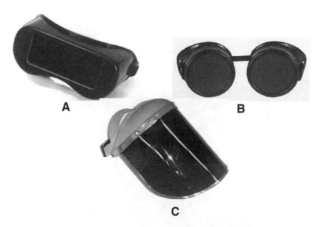

Figure 2-1. Safety glasses or goggles are necessary for protection during oxyfuel gas welding and cutting. A—One-piece goggles. B—Two-piece goggles. C—Face shield with tinted lens. (Uniweld Products, Inc.)

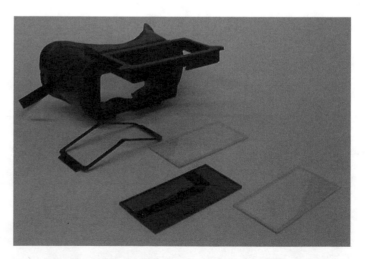

Figure 2-3. Clear cover lenses should be placed in front of and behind the colored filter lens.

Figure 2-5. The welder is wearing proper protective clothing, including safety goggles, gloves, coat, and cap.

Figure 2-4. Gloves are important in oxyacetylene welding. A—Fire-resistant leather gloves. B—Leather welding gloves with long cuffs for greater arm protection.

Protective clothing must be worn during welding and cutting. Trousers should not have cuffs, especially if the welder is cutting. Cuffs can catch burning metal particles. Outer clothing should be fire-resistant and free of grease or oil. Heavy cutting operations may require the use of a leather apron. High boots and leather leggings may also be needed. Some welding operations, particularly overhead welding, requires a cap. Wearing a cap when doing any type of welding or cutting is good practice.

Leather or fabric gloves are worn to protect the welder's hands and wrists, **Figure 2-4.** Gloves should extend to the end of the shirt or jacket sleeves. **Figure 2-5** shows a welder wearing proper protective clothing.

Welding Equipment

Worn hoses, faulty regulators, and poor hose or gauge connections can cause accidents. Check them carefully, and repair or replace them when necessary. Test all hose and regulator connections for leaks using a soapy solution. *Never use a flame to test for leaks!*

Make sure no oil or grease are used near oxygen hose fittings, cylinders, or torches. Oxygen in contact with oil or grease is a flammable combination.

Fuel gases should be referred to by their proper names, not just as "gases". Check cylinder labels to be sure of their contents. Acetylene, oxygen, and other cylinders containing high-pressure gases must be handled with care. Oxygen cylinders should be stored separately from acetylene or other fuel gas cylinders. Always store cylinders away from combustible materials. Keep safety caps in place. Fasten cylinders with chains or steel bands to a wall or post so they do not topple over. Cylinders should always be handled, stored, and used in an upright position. See **Figure 2-6.** A cylinder that is not properly secured is extremely dangerous.

Use a cylinder truck to move cylinders. If a cylinder truck is not available, the cylinder may be tipped and rolled along its bottom edge. Make sure the cylinder cap is on. The person moving the cylinder must use both hands and be knowledgeable of safety precautions.

The same precautions for storing a cylinder must be followed when using a cylinder in a welding area. *Never use acetylene at a pressure above 15 psig (103 kPa).* It becomes unstable and can cause an explosion. The safe use of fuel gases and welding equipment is discussed more in later units.

Figure 2-6. Properly stored gas cylinder chained to a wall.

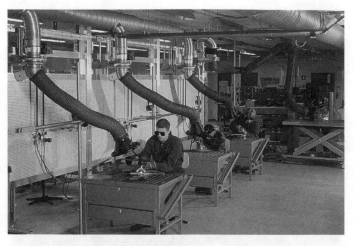

Figure 2-7. Fume extractors are used to remove fumes from the welding area. Proper ventilation is necessary to keep fresh air available. (Nederman, Inc.)

Ventilation

Fumes and gases generated during welding, particularly those from fluxes and metal alloys, can be poisonous. Some gases can replace oxygen in the air. Lack of oxygen causes a person to lose consciousness and can be fatal.

Proper ventilation is required to remove welding fumes and gases from the welding area. Opening doors and windows may be sufficient to remove contaminated air. A mechanical system can also be used, **Figure 2-7.** In some situations, a respirator or air purifier must be worn to provide fresh air to the welder, **Figure 2-8.** Welding must not be performed if proper ventilation is not available.

Safe Practices

Follow these safety precautions to prevent accidents and injury to yourself and others:

- Remove all flammable materials from the welding area before welding or cutting.
- Set up a screen or curtain to protect others from flying sparks or harmful rays.
- Never weld a completely closed container, such as a pipe capped on both ends. Expansion of air inside the container could cause an explosion.
- Never use oxygen as a substitute for compressed air.
- Treat every piece of metal as if it were hot until you are sure it is not. Use pliers or tongs to move hot metal or to cool hot metal in water. If you must leave a hot piece of metal in a welding area, mark it "hot" with a piece of chalk to warn others.

Check Your Progress

Write your answers in the spaces provided.

1. What is the most important personal protective device in the welding area?

_____

Figure 2-8. A small electrically powered air purifier supplies this welder with clean air.

2. What filter lens number should be worn when:
   a. oxyfuel gas welding material 1/8″ (3.2 mm) to 1/2″ (12.7 mm) thick?
   b. oxyfuel gas cutting material under 1″ (25.4 mm thick)?

   a. _____

   b. _____

3. Welding goggles with proper filter lenses will protect the welder's eyes from (circle letter):
   a. hot metal and sparks.
   b. injurious rays of light.
   c. both of the above.
   d. neither of the above.

4. Why must contact lenses *not* be worn in a welding area?

_____

_____

5. Why should gloves extend to the end of shirt or jacket sleeves? _____

_____

6. All hose and regulator connections should be checked for leaks with a(n) _____; *never* use a(n) _____.

7. Why should a completely closed container *never* be welded?

_____

_____

8. A properly stored cylinder should be secured with _____ or _____ to a wall or post.

9. Why should acetylene gas *never* be used over 15 psig (103 kPa)?

_____

_____

10. Why is proper ventilation during oxyfuel gas welding so important? _____

_____

_____

Instructor's Initials _____ Date _____

Things to Do

1. Check hose connections and regulator fittings for leaks using a soapy solution. If leaks are found, tighten the fitting, or replace it if necessary.

2. Clean the welding area. Throw away oily rags, paper, and other flammable materials.

3. Check welding goggles for broken or dirty lenses and worn or loose headbands. Clean and replace worn or broken parts as necessary.

4. Examine the oxygen and acetylene cylinders in storage areas. Make sure all cylinders are properly secured and empty cylinders are marked for return to the supplier. Use chalk to mark empty cylinders MT (empty) for easy identification. Empty cylinders should be separated from full cylinders.

5. Check the welding area for good ventilation. Doors and windows should open freely, and a ventilation system should be operating properly.

Unit 3

MEASUREMENT FOR WELDING

Systems of Measurement

Making accurate measurements for cutting and assembling pieces to be welded is an important part of a welder's job. The welder must be able to read a rule and use other measuring tools with precision, **Figure 3-1.** Typical tasks include taking measurements from a drawing, cutting pieces to exact lengths and angles, and setting up the pieces for welding.

Two measurement systems are used in the United States. One is the US conventional (English) system. The second is the SI metric measurement system, which is standard in most other countries. In the US conventional system, most measurements are made in inches. Each inch is divided into smaller units called graduations. Graduations are half of a larger unit, **Figure 3-2.**

The welder should be able to measure lengths to 1/32″ (inch) in the US conventional measure. In the metric system, measurements are made to the nearest millimeter (mm). Precision welding may require measurements even smaller than 1/32″ or 1 mm. Calipers, as shown in Figure 3-1, measure as small as 0.0005″ and 0.01 mm.

Using a Rule

Examine the edge of a rule, and count the number of graduations. Some rules will have the number 16 or 1/16 stamped along the face. This means there are 16 graduations per inch, and each graduation is 1/16 of an inch. Some rules may be stamped 1/32, 1/8, 10, or 100. This number indicates the smallest graduation on the edge of the rule. Graduation lines

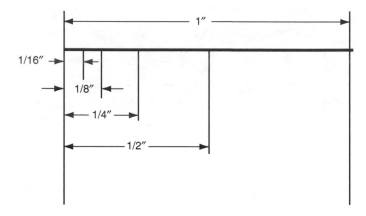

Figure 3-2. Each graduation in the US conventional system is half of the next larger unit of measurement. (Drawing not to scale.)

are different lengths, which makes it easier to locate units of measure, **Figure 3-3.**

A good way to practice using a rule is to measure metal pieces in the shop. Determine length, width, and thickness. Fractional measurements should be reduced to their lowest common terms. For example, a measurement of 12/16 should be reduced to 3/4. A measurement of 6/16 should be reduced to 3/8, and so on.

Units of Metric Measurement

Along the left edge of a metric rule, you will see the letters cm or mm, **Figure 3-4.** In some cases, the letters indicate the graduations are centimeters or millimeters. In other cases, the letters indicate the smallest graduations on the rule.

Metric graduations, or parts, are divided into units divisible by ten. A centimeter is one of 100 parts in a meter.

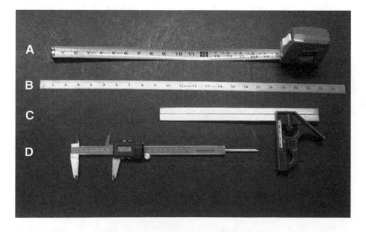

Figure 3-1. Measuring tools used by a welder. A—Tape measure. B—Steel rule. C—Combination square. D—Caliper.

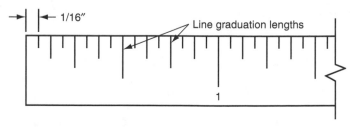

Figure 3-3. Graduation lines are of different lengths. The smaller the graduation, the shorter the line length.

A millimeter is one of 1000 parts in a meter and one of 10 parts in a centimeter.

In the SI metric system, dimensions are called out only as whole numbers, not fractions of whole numbers. For example, you would not say, one and a half centimeters; you would say 15 millimeters. The dimensions are the same but expressed as a whole number.

Other units of metric measure with which you should be familiar are:

- Kilogram, the metric unit of weight or mass. It is used like pounds.
- Pascal (Pa), the measure of pressure. It replaces pounds per square inch (psi).
- Temperatures given in degrees Celsius (°C) rather than degrees Fahrenheit (°F). In Celsius, water freezes at 0°C and boils at 100°C. By comparison, water freezes at 32°F and boils at 212°F.
- Volume, given in cubic centimeters or liters. It takes slightly more than 16 cubic centimeters (cm^3) to equal 1 cubic inch (in^3).

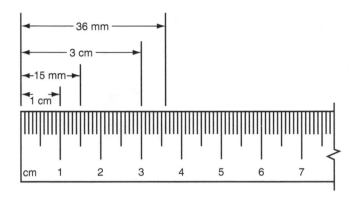

Figure. 3-4. A metric rule showing graduations in increments (increases) of 1 mm. The longest graduation marks are centimeter divisions.

Check Your Progress

Write your answers in the spaces provided.

1. Find the distance in inches from the left end of the rule to each graduation indicated by a blue line and letter.

A. _____

B. _____

C. _____

D. _____

E. _____

F. _____

G. _____

H. _____

I. _____

J. _____

K. _____

L. _____

M. _____

N. _____

O. _____

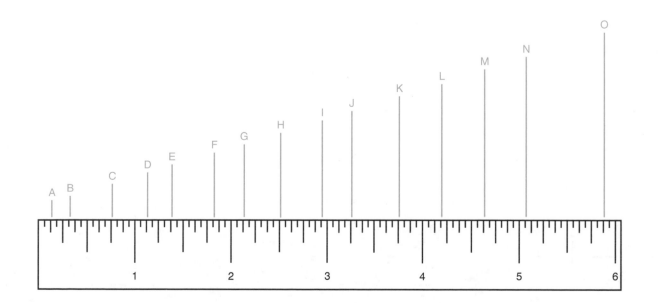

2. The welder is often required to measure metal parts. Measure these pieces with a rule graduated in the US conventional system.

A. _____   H. _____   N. _____   T. _____

B. _____   I. _____   O. _____   U. _____

C. _____   J. _____   P. _____   V. _____

D. _____   K. _____   Q. _____   W. _____

E. _____   L. _____   R. _____   X. _____

F. _____   M. _____   S. _____   Y. _____

G. _____

3. Find the distance in millimeters from the left end of the rule to each graduation indicated by a blue line and letter.

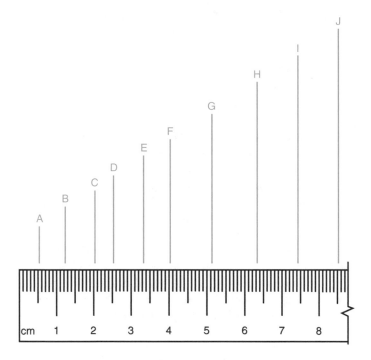

A. _____

B. _____

C. _____

D. _____

E. _____

F. _____

G. _____

H. _____

I. _____

J. _____

4. Measure the pieces in Problem 2 with a rule graduated in the SI metric system.

A. _____

B. _____

C. _____

D. _____

E. _____

F. _____

G. _____

H. _____

I. _____

J. _____

K. _____

L. _____

M. _____

N. _____

O. _____

P. _____

Q. _____

R. _____

S. _____

T. _____

U. _____

V. _____

W. _____

X. _____

Y. _____

Instructor's Initials _____ Date _____

Things to Do

1. Study the metric and US conventional conversion tables in the back of this book.

2. Look over the various measuring tools in your shop to determine if they use the metric or US conventional system of measurement.

3. With your instructor, study the various measurements shown on welding plans.

Unit 4

WELDING SYMBOLS

Welding Symbols on Drawings

Welding requires the welder be skilled in reading mechanical drawings. A mechanical drawing describes exactly how the parts will be assembled. The welding information is conveyed using standard *welding symbols.* Symbols tell the welder the location and size of the weld, type of joint to be used, and other important information.

Welding symbols are standardized by the American National Standards Institute (ANSI) and the American Welding Society (AWS). *ANSI/AWS A2.4 Standard Symbols for Welding, Brazing, and Nondestructive Examination* is the standard for welding symbols.

How to Read Welding Symbols

A welding symbol is created and read from the reference line. The *reference line* is a horizontal line around which elements are placed. An arrow at one end of the reference line points to the place on the drawing where the desired weld is to be made. The arrow line will always be at an angle to the reference line.

Weld symbols are used to describe a desired weld, **Figure 4-1.** A weld symbol is placed above or below the reference line. A weld symbol placed *below* the reference line refers to the arrow side of the joint (the side to which the arrow points). See **Figure 4-2.** A weld symbol placed *above* the reference line refers to the other side of the joint (the side opposite the arrow). See **Figure 4-3.** Standard locations of elements on the welding symbol are shown in **Figure 4-4.** A tail is needed only if information is placed in the tail.

If it is necessary to indicate the exact size of a weld, place the dimensions on the welding symbol as shown in **Figure 4-5.**

Additional information can be added to a welding symbol to completely describe the joint. Refer to Figure 4-4. Groove angle (abbreviation A) and root opening (abbreviation R) may be indicated.

The size or strength of a weld (abbreviation S), the length and pitch (abbreviation L-P) of an intermittent or skip weld, and the contour of the weld bead may also be part of the welding symbol. The number of spot welds is shown as (N). If a weld is to be continuous around some feature, a weld-all-around symbol is added.

The field weld symbol indicates the welding operation will be done at the installation or construction site. See **Figure 4-6.** A weld with a field weld symbol will not be made in the shop. Study the typical welding symbols in **Figures 4-7** and **4-8.**

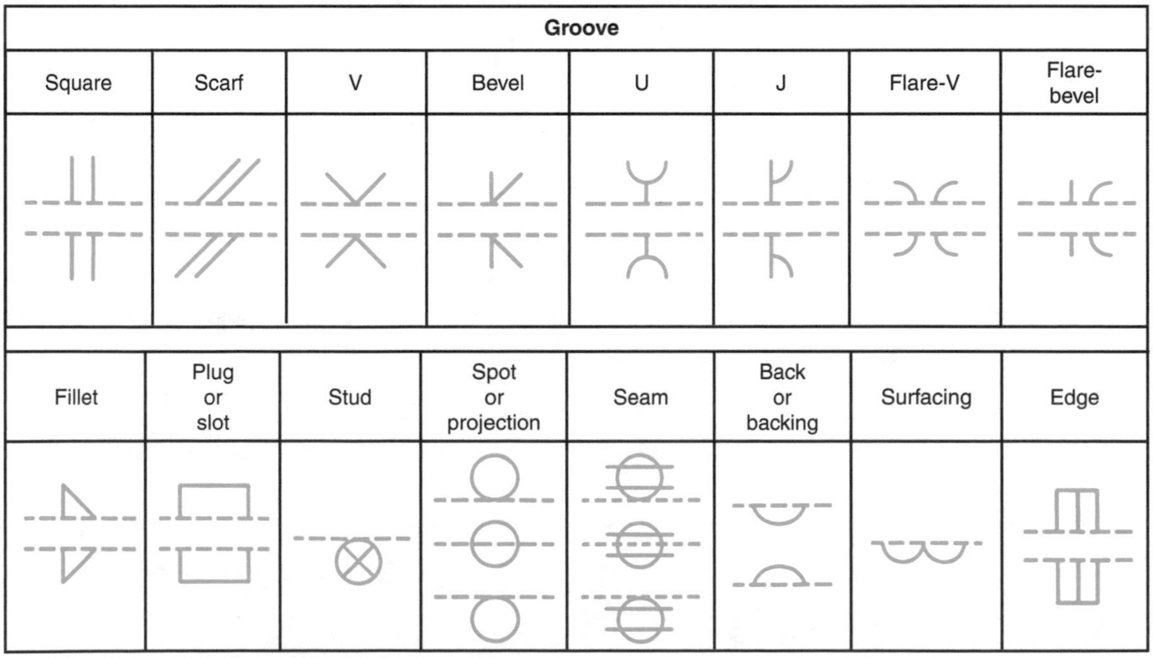

Groove							
Square	Scarf	V	Bevel	U	J	Flare-V	Flare-bevel

Fillet	Plug or slot	Stud	Spot or projection	Seam	Back or backing	Surfacing	Edge

Note: The reference line is shown dashed for illustration purposes.

Figure 4-1. Weld symbols. (From *ANSI/AWS A2.4-98*. Printed with Permission of the American Welding Society)

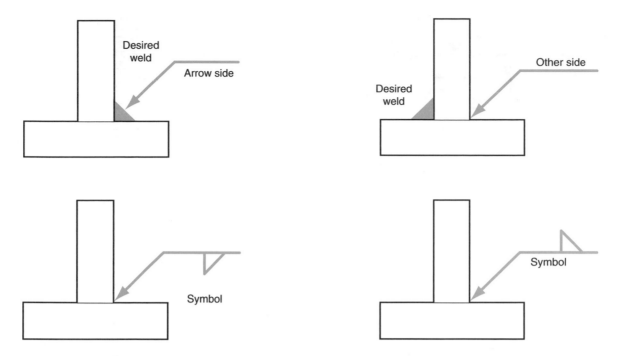

Figure 4-2. The weld symbol is below the reference line, indicating the weld is to be made on the arrow side of the joint.

Figure 4-3. The weld symbol is above the reference line, indicating the weld is to be made on the side of the joint opposite the arrow.

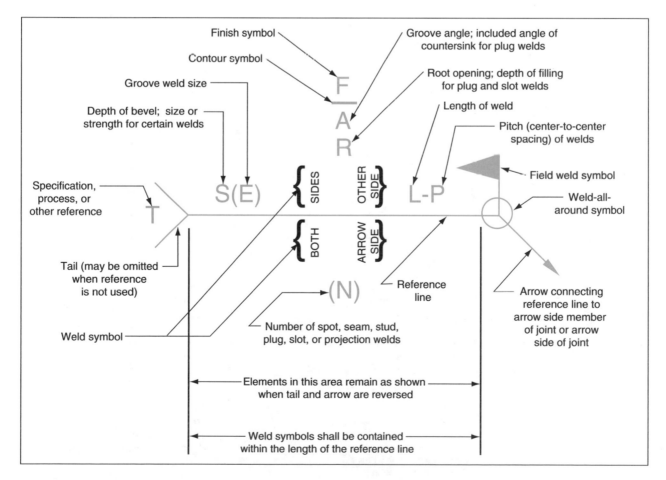

Figure 4-4. Standard locations of elements on the American Welding Society welding symbol. (From *ANSI/AWS A2.4-98.* Printed with Permission of the American Welding Society)

Figure 4-5. Dimensions are used with the symbol to describe the size of the weld.

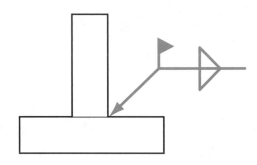

Figure 4-6. Use of the field weld symbol.

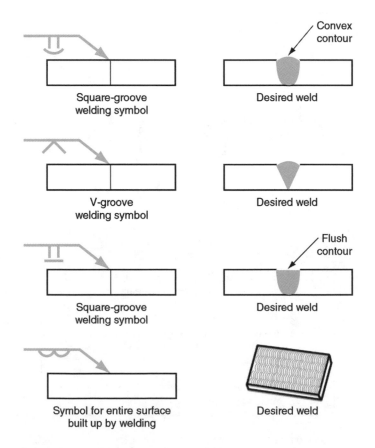

Figure 4-7. Some typical welding symbols and their meanings.

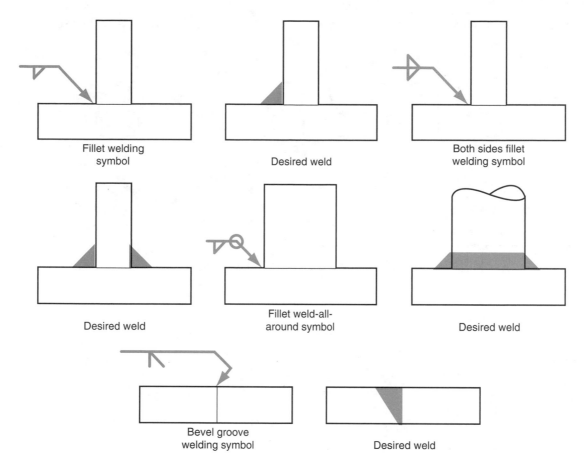

Figure 4-8. More welding symbols and their meanings.

Check Your Progress

Follow the directions given for each problem.

1. Add information to the reference lines and arrows to completely describe the joints indicated.

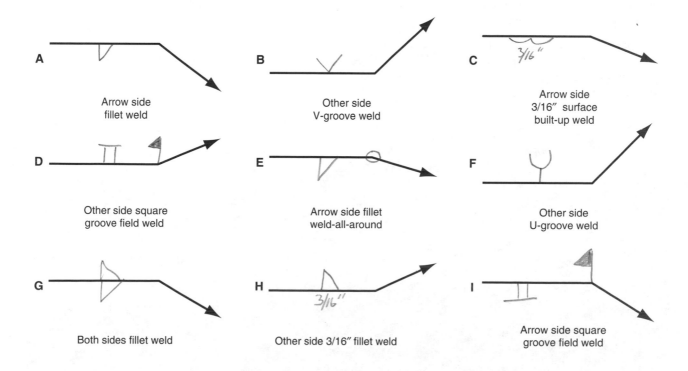

A — Arrow side fillet weld

B — Other side V-groove weld

C — Arrow side 3/16″ surface built-up weld

D — Other side square groove field weld

E — Arrow side fillet weld-all-around

F — Other side U-groove weld

G — Both sides fillet weld

H — Other side 3/16″ fillet weld

I — Arrow side square groove field weld

2. Each item has a set of two drawings. Read the symbol on the left-hand drawing. Then, draw the weld on the right-hand drawing.

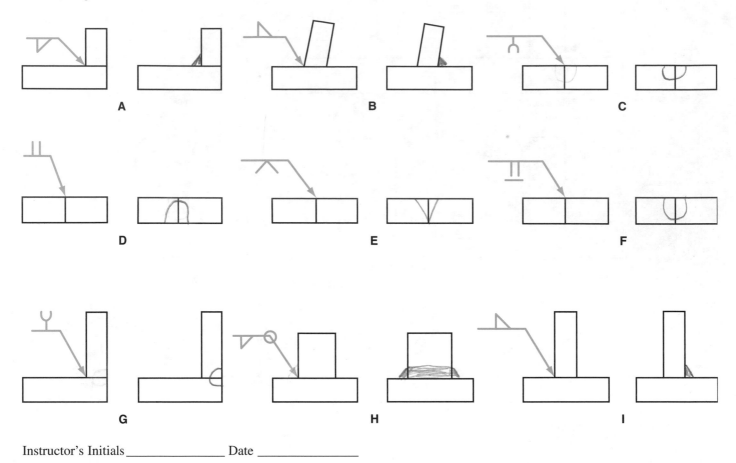

Instructor's Initials _____ Date _____

Things to Do

1. Locate a metal part that has some portions assembled by welding. Make a drawing of the part, and show the correct welding symbol for each weld.

2. Ask your instructor for a drawing of a metal part that requires a number of welds. For each welding symbol shown on the drawing, make a sketch of how the actual weld would look when completed.

3. Using Figure 4-4 as a reference, draw a welding symbol that includes the following information:
   a. Square groove weld.
   b. Flush with the surface.
   c. To be made in the field.

Unit 5

PREPARING THE JOINT

Types of Joints

A *weld joint* refers to the place or area where the ends, surfaces, or edges of two pieces of metal are to be joined with a weld. There are five basic types of weld joints, as shown in **Figure 5-1**:

- Lap joint.
- Butt joint.
- T-joint.
- Corner joint.
- Edge joint.

A *lap joint* combines two overlapping pieces of metal. The bottom surface of one piece overlaps the top surface of another piece. A *butt joint* is used to combine two pieces of metal end to end or side to side. A *T-joint* is formed when two pieces of metal are joined at an angle of or near 90° to one another. The edge of one piece is joined to the surface of the other piece at any point except the edge. A *corner joint* is formed when the edge of one piece of metal is joined to the edge of another piece at an angle of approximately 90°. The *edge joint* is formed when all or part of the surface areas of two pieces of metal are in contact and at least two of the edges are flush. The pieces are joined by welding along the flush edges.

Selecting and Preparing Joints

Three steps must be taken to prepare two pieces of metal for joining. First, the type of weld joint must be selected. Generally, the simplest joint that best serves the requirements of the job is selected. Second, the weld joint must be prepared for welding. The metal pieces need to be measured and cut to appropriate size. In some cases, the pieces might need to be machined, beveled, or bent. Third, the surfaces and edges of the joint must be cleaned.

Properly selecting and preparing the joint will help produce the strongest weld possible. Weld joints are prepared in order to give deep, complete penetration while keeping the metal edges true and straight.

Variations in joint types and preparation depend on metal thickness. Figure 5-1 shows the joints without any edge preparation. This is typical for welding sheet metal, usually 1/8″ (3.2 mm) or less in thickness. These same joints are also used with thicker plates. However, the edges of thicker material must be prepared using such processes as flame cutting, machining, or grinding to enable full penetration of the weld. **Figure 5-2** shows a prepared butt and corner joint.

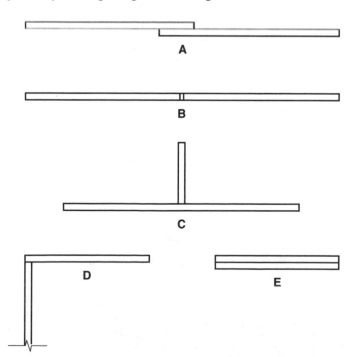

Figure 5-1. Joints used in welding. A—Lap joint. B—Butt joint. C—T-joint. D—Corner joint. E—Edge joint. All edges should be straight and smooth.

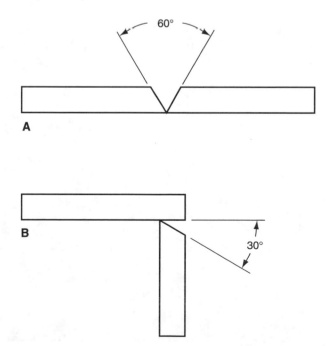

Figure 5-2. Joint preparation for welding thick steel plate. Angles may be increased with metal thickness. A—Beveled butt joint. B—Outside corner joint.

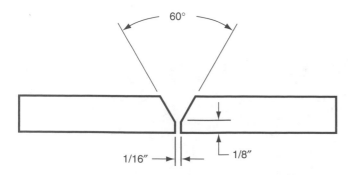

Figure 5-3. Preparation of a single V-groove joint on thick steel plate.

When preparing joints on thin sheet metal, be sure the edges have been cut straight with shears. If the metal edges are curled or bent, tap them with a mallet to flatten and straighten them. Most sheet metal joints need no further edge preparation whether using filler metal or not.

Steel plate, 1/8″ to 3/8″ (3.2 mm to 9.5 mm) thick, can be cut easily with a power hacksaw or bandsaw. In some cases, it is necessary to cut thick steel with an oxyacetylene torch (as discussed in Unit 22).

To obtain a full-penetration weld on thick steel plate, a single V-groove must be prepared, **Figure 5-3.** Always wear safety goggles when grinding. See **Figure 5-4.**

Figure 5-4. The operator is using a pedestal grinder to grind a 30° bevel on a steel plate. Notice he is wearing proper protective equipment.

Figure 5-5. Wire brushing an outside corner joint to remove all contaminants is a necessary step in preparing the joint.

Cleaning the Metal

Metal joints should be clean and dry before welding. All joints should be cleaned with a stainless steel brush, steel wool, abrasive paper, or a grinder. The metal pieces are cleaned to remove all surface contaminants, such as rust (oxidation), paint, dirt, or oil. See **Figure 5-5.** The strength of the weld may be greatly reduced if these contaminants are not removed from the joint.

Grease and paint can usually be removed with strong solvents. Solvents are flammable; always use them away from the welding area. Dry the metal completely before attempting to weld. A portable grinder or disc sander may be used to remove rust and paint. Care should be taken to keep the edges straight and the surfaces smooth when grinding.

Check Your Progress

Write your answers in the spaces provided.

1. The first two steps to consider when preparing to weld are:

   a. _THE TYPE OF JOINT_

   b. _THE JOINT MUST BE PREPARED_

2. The five basic joints used in welding are:

   a. _BUTT_

   b. _EDGE_

   c. _CORNER_

   d. _LAP_

   e. _TEE_

3. The thickness of sheet metal, when no joint preparation is required to obtain full penetration, is *(circle letter)*:
   a. 1/32″ (0.8 mm) or less.
   b. 1/16″ (1.6 mm) or less.
   c. 1/8″ (3.2 mm) or less.
   d. 3/16″ (4.8 mm) or less.

4. Steel plate can be easily cut with _HACKSAW OR BAND SAW_.

5. Two methods of cleaning metal before welding are:
   a. _Grinder_
   b. _SS Brush_

Instructor's Initials _____ Date _____

Things to Do

1. Using scrap steel plate 1/8″ to 1/4″ (3.2 mm to 6.4 mm) thick, practice grinding the edges to prepare them for a single V-groove joint.

2. Practice cutting 1″ (25 mm) wide strips of sheet steel with a hand shear. After cutting, lightly hammer the strips with a mallet on a flat table to straighten the edges.

3. If your shop has rusted steel plate used for practice welding, grind or disc sand the plate surface in preparation for welding.

4. In the space below, make sketches of a single V-groove and a corner joint for 1/4″ (6.4 mm) thick steel plate.

OXYFUEL GASES AND CYLINDERS

Oxygen and Acetylene

A number of different fuel gases can be combined with oxygen under controlled conditions to create a flame for welding. Each oxygen/fuel gas combination creates a flame with a different temperature, **Figure 6-1.** A combination of oxygen and acetylene generates a flame with the highest temperature. You must understand the nature of these two primary gases to be able to use them safely and skillfully. Other fuel gases are used for welding and cutting, but this textbook concentrates on oxygen and acetylene since this is the most widely used combination for industrial processes.

Oxygen

Oxygen (chemical symbol O) is one of the most common elements on earth. It is a colorless, odorless, tasteless gas. In its gaseous state, oxygen exists as a molecule composed of two oxygen atoms. It is designated as O_2.

Air contains about 21% oxygen and provides a readily available source of oxygen for commercial production. Oxygen is obtained by cooling air to a very low temperature until it becomes a liquid. Then, it is separated from other elements in the liquid by distillation. *Distillation* is a process of raising the temperature to a certain point and allowing a desired element, like oxygen, to boil out of the liquid. The resulting oxygen used

for welding is about 99.5% pure. Under normal conditions, 12.07 cu. ft. (0.34 cu. m) of oxygen weighs 1 lb. (0.45 kg). It is just slightly heavier than air.

The most important property of oxygen for welding purposes is its ability to support combustion. Oxygen will not burn alone, but when combined with most other elements, the compound becomes highly flammable. When oxygen is combined with acetylene, the mixture produces one of the hottest gas flames obtainable, with a temperature of 5589°F (3087°C).

Combustibility makes oxygen one of the most valuable sources of heat for welding; however, it is also a hazard if not handled carefully. Since oxygen supports combustion, it must never be used like compressed air to blow off dust or dirt from clothing or equipment. Oil should never be used around oxygen cylinders, regulators, hoses or torches, since an explosion may result. Oxygen should only be used in welding equipment with proper controls.

Acetylene

Acetylene (C_2H_2) is a chemical compound composed of equal parts of carbon and hydrogen. Acetylene is made by bringing calcium carbide into contact with water. It is a colorless gas with a pungent odor. Acetylene is very unstable and can become highly explosive when mixed with oxygen or air.

Under pressure, acetylene will dissociate, or break up, into its component parts. This dissociation releases energy in the form of heat, which makes acetylene gas so useful. However, when subjected to pressures above 15 psig (103 kPa), *it can cause a violent explosion.* Acetylene gas should never be compressed above 15 psig (103 kPa) in its free state in cylinders, pipes, hoses, regulators, or torches.

Note: *Psig* or *psi* means pounds per square inch gauge. It is the unit of measurement for pressure in the US conventional system. The metric measure of pressure is given in pascals (Pa) and expressed in thousands of pascals or *kilopascals* (kPa). One pound per square inch is equal to 6.8948 kilopascals (kPa). To convert psig to kPa, use this formula: *kPa = number of pounds per sq. in. × 6.8948.* For example, to find the number of kPa equal to 15 psig:

kPa = 15 × 6.8948
kPa = 103.4

Storing Oxygen and Fuel Gases

Knowing the chemical properties of oxygen and acetylene is helpful to understand the great potential, as well as danger, inherent in these gases. To be safely and productively used for

Fuel Gas	Temperature of Neutral Flame	
	°F	°C
Acetylene	5589	3087
Hydrogen	4820	2660
Methylacetylene –Propadiene	5301	2927
Natural Gas (methane)	4600	2538
Propane	4579	2526
Propylene	5250	2900

Figure 6-1. Fuel gases used for oxyfuel gas welding and cutting, and their neutral flame temperatures when combined with oxygen.

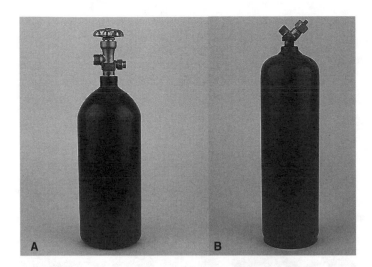

Figure 6-2. Gas storage cylinders. A—An oxygen cylinder contains 2200 psig (15 170 kPa) of compressed oxygen. B—An acetylene cylinder contains 200 psig to 300 psig (1380 kPa to 2070 kPa) of compressed acetylene gas.

welding, the gases must first be compressed and stored in special containers called *cylinders.* The physical properties of oxygen and acetylene are quite different; therefore, different types of cylinders must be used for each gas.

Oxygen and acetylene cylinders are built to rigid specifications, **Figure 6-2.** They must be extremely strong and durable to safely handle high-pressure compressed gases. Care must be taken in storing, transporting, and handling these cylinders. When handled properly, oxygen and acetylene cylinders are very safe.

Oxygen Cylinders

Oxygen cylinders are made from seamless drawn steel and shaped by dies into tanks. They provide strong, durable containers with no joints or welded seams. The thickness of the cylinder wall is at least 1/4″ (6.4 mm). **Figure 6-3** shows the construction details of a typical oxygen cylinder.

When fully charged, a standard oxygen cylinder holds about 244 cu. ft. (7 cu. m) of oxygen at a pressure of 2200 psig (15 169 kPa) at 70°F (21°C). At this pressure, the cylinder has the explosive force of a bomb! If it were dropped or hit with a heavy object, or if the valve were knocked off, it could explode and cause severe damage. Therefore, proper handling of an oxygen cylinder is mandatory. (See the description of cylinder care.)

The only opening in an oxygen cylinder is a threaded hole at the top where the valve is fitted. Specially designed bronze valves are used to contain the high pressures of oxygen cylinders, **Figure 6-4.** When in use, the oxygen cylinder valve must be completely open to prevent the escape of oxygen around the valve stem. The valve should be tightly closed when the cylinder is not in use. Oxygen cylinders also have a threaded safety cap which protects the valve from being damaged or broken off. Replace the safety cap each time the cylinder is transported or not in use.

Acetylene Cylinders

Acetylene cylinders are made from welded steel tubes filled with a highly porous material that acts like a sponge. The

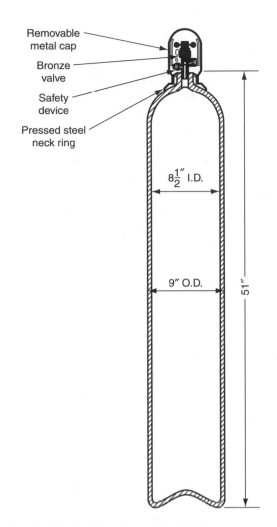

Figure 6-3. Details of a 244 cu. ft. oxygen cylinder.

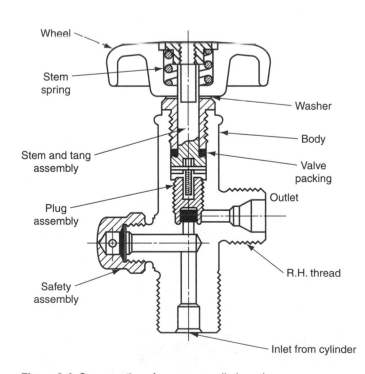

Figure 6-4. Cross section of an oxygen cylinder valve. (Airco Welding Products Div.)

material is saturated with liquid acetone. Acetone has the ability to absorb acetylene gas much like cotton absorbs water. The acetylene gas can be safely compressed in the cylinder at pressures between 200 psig and 300 psig (1380 kPa and 2070 kPa). **Remember:** Acetylene gas must be released from the cylinder at working pressures below 15 psig (103 kPa).

Construction details of an acetylene cylinder are shown in **Figure 6-5.** Filler materials such as charcoal, asbestos, or balsa wood are used to absorb the acetone. Low-temperature melting fuse plugs at the top and bottom of the cylinder allow gas to escape if temperatures exceed 212°F (100°C). Escaping gas may cause a fire, but no explosion will occur.

A steel valve is fastened to the top of the acetylene cylinder. A special key is used to open and close the valve. Opening the valve 1/4- to 1/2-turn will produce sufficient gas flow. In case of an emergency, the cylinder valve can be quickly shut off. The key must always be in place when the cylinder is in use.

A typical acetylene cylinder contains about 280 cu. ft. (8 cu. m) of gas at a pressure of 250 psig (1724 kPa). The threaded metal safety cap must be in place when the cylinder is not in use. The safety cap protects the steel valve from being damaged or broken off.

Cylinder Care

Follow these rules for the safe handling and storage of oxygen and acetylene cylinders:

- Handle, store, and use cylinders in an upright position. Secure each cylinder to a wall or post using chains or straps.
- When transporting cylinders, use a properly designed cylinder truck. Chain or strap the cylinder to the truck to prevent it from tipping or falling over, **Figure 6-6.**

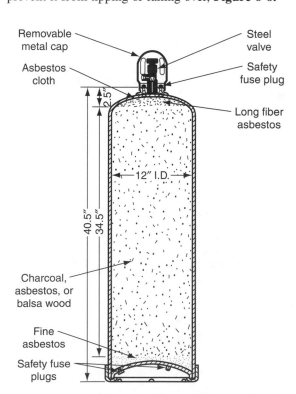

Figure 6-5. Details of a typical acetylene cylinder. (Airco Welding Products Div.)

Labels: Removable metal cap, Asbestos cloth, Steel valve, Safety fuse plug, Long fiber asbestos, 2.5", 12" I.D., 40.5", 34.5", Charcoal, asbestos, or balsa wood, Fine asbestos, Safety fuse plugs

Figure 6-6. Oxygen and acetylene cylinders should be securely attached with chains or straps to a welding cart.

- Replace cylinder safety caps after each use to protect the valves.
- Use hand pressure to open oxygen cylinder valves. *Never use a wrench!* Open acetylene cylinder valves using the key supplied.
- Never roll cylinders horizontally or use them as rollers to move a load.
- Store oxygen and acetylene cylinders separately, at least 20 ft. (6.1 m) from each other and any combustible material.
- Do not store cylinders in areas with hot temperatures as internal pressure could rise to a dangerous level.
- Store full cylinders away from empty cylinders.
- Fully open the oxygen cylinder valve when in use so oxygen cannot leak around the valve stem.
- Open the acetylene cylinder valve only 1/4- to 1/2-turn so it can be quickly shut off in an emergency.

Check Your Progress

Write your answers in the spaces provided.

1. Oxygen is a gas that is _____, _____, and _____.

2. What is the most important property of oxygen for welding purposes? _____

_____

_____

3. Why should oil never be used around oxygen equipment, such as regulators or torches? _____

_____

_____

4. List the two elements that make up acetylene gas.

a. _____

b. _____

5. Acetylene is made by bringing _____ into contact with _____.

6. A violent explosion may result if acetylene gas is subjected to working pressures above _____.

7. Acetylene cylinders are made from (circle letter):
   a. die-stamped steel sheet.
   b. seamless drawn steel.
   c. welded steel tubes.
   d. threaded pipe sections.

8. Specially designed _____ made of _____ are used to contain the high pressures in oxygen cylinders.

9. When should a threaded safety cap be used on acetylene and oxygen cylinders, and why? _____

   _____

   _____

   _____

10. List the filler materials used to absorb the acetone in acetylene cylinders.

    a. _____

    b. _____

    c. _____

Instructor's Initials _____ Date _____

Things to Do

1. Under instructor supervision, smell a very small quantity of acetylene gas. Try to describe the odor in your own words. Learn to recognize the odor to detect leaks in welding equipment, hoses, and cylinders.

2. Write a report on the production of acetylene gas for welding. Ask your instructor for sources of information or do library research.

3. Prepare a class presentation on the topic, "How Oxygen is Obtained from Liquefied Air."

4. Explain how you can determine oxygen is slightly heavier than air. You may want to ask a chemistry teacher for help.

5. Check the safety precautions discussed in this unit, and see if any potentially dangerous situations exist in your welding area. If so, talk with your instructor and discuss changes to correct the dangers.

6. Visit a shop that does welding, and observe the types of acetylene and oxygen cylinders being used. If some cylinders are different from those in your shop, ask why. Report your findings to the class.

7. Write to manufacturers of acetylene and oxygen cylinders to ask for their catalogs. Prepare a list of the different sizes of cylinders being produced according to their volume.

8. Make arrangements for the storage of acetylene and oxygen cylinders that cannot be properly stored now.

Unit 7

WELDING TORCHES AND TIPS

Functions of Torches

Welding torches have two primary functions. The first is to control and mix the oxygen and fuel gases within the mixing chamber of the torch. The second is to direct the flame to the welding, brazing, or soldering work area.

The experienced welder is skilled at overseeing these two functions. First, by properly adjusting the hand valves on the torch, the welder is able to *throttle* (make the final flow adjustment to) the gases being fed to the torch. This enables the welder to produce the proper flame for the job being performed. Second, by correctly handling the torch and using established welding techniques, the welder can apply and direct the necessary amount of heat to the workpiece.

The welding torch consists of five main parts, **Figure 7-1**:
- Torch body or handle.
- Oxygen and fuel gas hose connections.
- Valves for regulating gas flow (also called hand valves).
- Mixing chamber.
- Welding tip.

Note: An important device that attaches to each of the hose connections is the flashback arrestor (discussed on page 30). The flashback arrestor is considered part of the welding torch. Some welding torches come with built-in flashback arrestors.

Torch Types

Two types of torches are commonly used for oxyfuel gas welding: the *positive pressure–type torch* (also known as an equal pressure–type torch or medium pressure–type torch), and

the *injector-type torch.* In both of these torch types, oxygen and fuel gas are combined in a mixing chamber. The mixed gases flow to the tip of the torch where they are burned.

The two types of torches differ primarily in the way oxygen and fuel gas are fed into the mixing chamber. The positive pressure–type torch feeds oxygen and fuel gas into the mixing chamber at relatively equal pressures. The injector-type torch uses high-pressure oxygen and low-pressure fuel gas.

The mixing chamber may be found within the torch body itself, or it may be part of the torch tip. If the mixing chamber is part of the torch tip, this entire assembly is changed when changing tip sizes.

Positive Pressure–type Torch

The positive pressure–type torch requires medium oxygen and fuel gas pressures to force the gases into the mixing chamber. The schematic drawing in **Figure 7-2** shows the passageways for oxygen and fuel gas to meet in the mixing chamber.

Fuel gas pressures must be above 0.25 psig (1.7 kPa) for the positive pressure–type torch to function. Normal operating pressures for both the fuel gas and the oxygen are between 1 psig and 15 psig (7 kPa and 103 kPa). In most welding processes, the oxygen and fuel gas pressures should remain relatively equal. This enables both gases to be evenly forced into the mixing chamber and properly mixed.

Injector-type Torch

The exterior of an injector-type torch (also called a low-pressure torch) looks like a positive pressure–type torch.

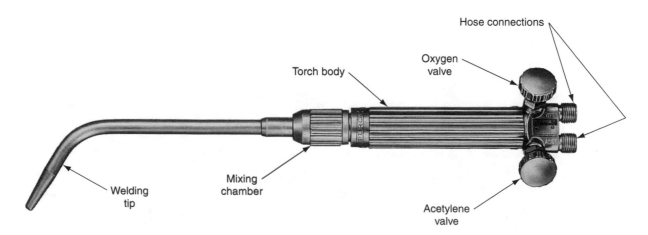

Figure 7-1. Exterior view of a typical welding torch.

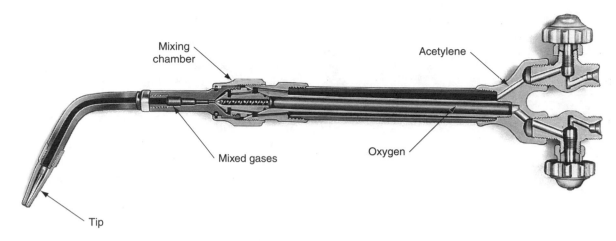

Figure 7-2. Cutaway view of a medium pressure torch. (Victor Equipment Co.)

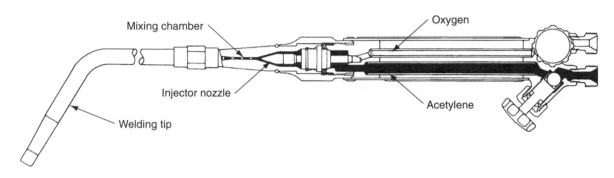

Figure 7-3. Cross section of an injector-type torch. (Linde Co. Div. of Union)

However, their internal constructions differ. The injector-type torch forces high-pressure oxygen through an injector, which simultaneously draws or pulls the low-pressure fuel gas into the mixing chamber. See **Figure 7-3.** The oxygen and fuel gas mix and flow to the tip where the proportions of the two gases remain constant.

One advantage of the injector-type torch is its ability to operate using very low acetylene or fuel-gas pressure. This enables it to be used with a low-pressure (0.25 psig or 1.7 kPa) acetylene generator. Another advantage is that it more completely draws out all the gas from an acetylene cylinder.

Torch Valves

Torch valves are located at either the hose connection end or the tip connection end of the torch body, **Figure 7-4.** One valve controls the flow of oxygen, and the other controls the flow of fuel gas into the torch. If the letters *FUEL*, *GAS*, *ACET*, or *F* appear on the torch body near the valve, the valve controls the flow of fuel. If the letters *OXY* or *O* appear on the torch body near the valve, the valve controls the flow of oxygen. The valves are used to turn on the gas flow, adjust the flame, and shut off the gas being fed to the torch. Torch valves are opened in a counter-clockwise direction and closed in a clockwise direction. Only finger force is needed to open and shut the valves (never use a wrench). Torch valves are also appropriately called hand valves. One-half to one full turn counterclockwise will open the valves completely.

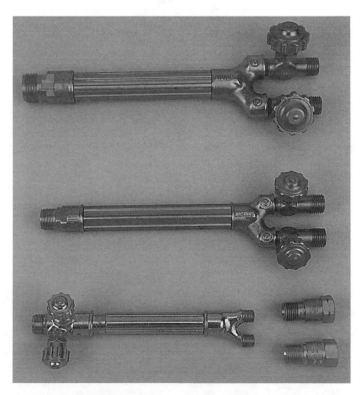

Figure 7-4. Torch bodies come in a variety of sizes and styles. Torch valves can be located at either end of the torch body. (Victor Equipment Co.)

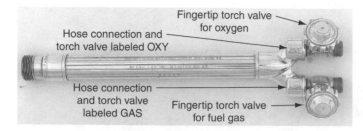

Figure 7-5. Torch valves and hose connections are clearly marked OXY for oxygen and FUEL or GAS for fuel gas. (Uniweld Products, Inc.)

Flashback Arrestors

A remote but real danger in oxyfuel gas welding is the occurrence of flashback. *Flashback* happens when the welding flame moves back into the mixing chamber or even farther to the hoses, regulators, and cylinders. A *flashback arrestor* is designed to prevent an explosion in the regulator or cylinder, **Figure 7-6.** The flashback arrestor is installed between the hose and the torch. It has a built-in check valve that prevents the flame from moving from the torch to the hose or regulator. Flashback is less likely to occur in a positive pressure torch.

Figure 7-6. The flashback arrestor is installed between the hose and the torch. It is designed to prevent an explosion in the regulator or cylinder. (Uniweld Products, Inc.)

Welding Tips

The oxyfuel gas welding flame is ignited and maintained at the end of a solid copper torch tip. Several types of welding torch tips are available, **Figure 7-7.** A welder may choose a one-piece tube-and-tip combination or a two-piece tip. The two-piece tip consists of a shortened torch tip threaded into a longer torch tube. Both the one- and two-piece torch tips are threaded onto the torch body where they are attached to the mixing chamber. Assembled torch tips, each with its own mixing chamber, are also available. These tips are often called *in-tip mixers*, **Figure 7-8.**

Copper conducts heat rapidly, so there is little danger of overheating or backfiring. However, a backfire may still happen. A *backfire* is a sharp popping sound that occurs when the welding tip is dirty or comes too close to the work. Following a backfire, the flame may continue to burn or it may be extinguished.

Tip Size and Heat Transfer

Tip size is measured by the diameter of the orifice. The *orifice* is a precise hole machine-drilled into the flame end of the tip. A number stamped on the tip indicates the size of the orifice. Generally, a small tip number indicates a small orifice.

The larger the tip size, the greater the gas flow through the orifice, and the greater the amount of heat produced. A small orifice will deliver a small amount of heat to the welding surface of the base metal, while a large orifice will deliver more intense heat.

Welding flame temperatures remain constant and do not change with tip size. Only the amount of heat supplied to the welding area changes. The welder controls the amount of heat by selecting the correct torch tip for each welding job and applying the proper techniques.

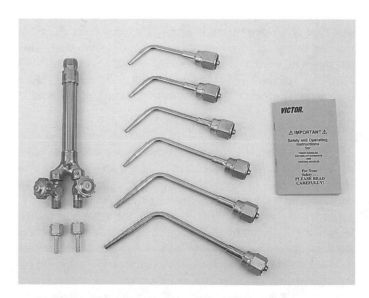

Figure 7-7. A torch body and an assortment of welding tips. Manufacturers always provide operating instructions and recommendation charts to guide the use of their welding equipment. (Victor Equipment Co.)

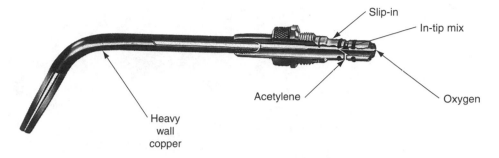

Figure 7-8. Cutaway view of a welding tip with an in-tip mixing chamber. (Smith Welding Equipment Co.)

Welding Tip Selection

To make good welds, it is necessary to pick the proper size tip. Different metal thicknesses require different amounts of heat to produce a quality weld. For example, a greater amount of heat is needed to weld thick steel than to weld thin steel. Thus, a larger tip size is required. If the torch tip orifice is too small, the heat will be inadequate to bring the metal to its melting and flowing temperatures. If the tip is too large, the excess heat will result in poor welds that are made too fast.

Follow the manufacturer's recommendation for the correct tip size for a welding job. Welding supply companies provide tip size recommendation charts to their customers. **Figure 7-9** shows a tip size chart for various thicknesses of metal. Unfortunately, tip numbering systems differ among tip manufacturers. No standard identification system has been established.

The size of an orifice can be found by using a set of drill bits. Be careful not to scratch or distort the orifice. Drill bits are numbered 1 through 80. The diameter of a Number 1 drill bit is 0.2280″ (5.79 mm). The diameter of a Number 80 drill bit is 0.0135″ (0.34 mm). **Note:** The larger the number of the drill bit, the smaller the diameter.

Figure 7-10. Standard set of tip cleaners with burnisher.

Care Using Welding Tips

Molten metal, slag, carbon, and dirt collect on the end of the torch tip during oxyfuel gas welding processes. These unwanted deposits cause the cone of the flame to be distorted rather than pointed. They can also cause a backfire.

The flat end of the tip and the orifice must be regularly cleaned to remove these deposits. A standard set of tip cleaners or a drill bit can be used to clean the orifice, **Figure 7-10.** The tip cleaner number and corresponding orifice size are listed in **Figure 7-11.** Care must be taken not to enlarge the orifice. The end of the tip can be polished smooth with a burnisher or a very fine mill file if it becomes scratched or rough, **Figure 7-12.** After filing the tip flat, the orifice may need to be cleaned again.

The welding tip is a delicate instrument, subjected to mechanical wear and flame erosion with each use. The following suggestions will help keep torch tips in good condition for the maximum amount of time:

- Use only tips designed for the torch being used. Do not interchange torch tips and bodies made by different manufacturers if they are not designed to work together.
- Use the correct size tip cleaner.
- Use a box-end torch wrench to minimize damage to the torch nut. *Never use pliers.*
- Do not try to remove a hot tip from a tip tube. Allow the tip and tube to cool first. Do not install a cold tip in a hot tip tube.
- Hang up the torch when it is not in use to prevent the tip from becoming scratched or damaged.

Welding Thickness (inches)	Welding Tip Orifice Size*	Welding Rod Diameter (inches)	Oxygen		Acetylene	
			Pressure (psig)	cfh	Pressure (psig)	cfh
1/32	74	1/16	1	1.1	1	1
1/16	69	1/16	1	2.2	1	2
3/32	64	1/16 or 3/32	2	5.5	2	5
1/8	57	3/32 or 1/8	3	9.9	3	9
3/16	55	1/8	4	17.6	4	16
1/4	52	1/8 or 3/16	5	27.5	5	25
5/16	49	1/8 or 3/16	6	33.0	6	30
3/8	45	3/16	7	44.0	7	40
1/2	42	3/16	7	66.0	7	60

*Note: The tip orifice size as shown is the number drill size. These recommendations are approximate. The torch maufacturers' recommendations should be carefully followed.

Figure 7-9. A typical welding chart used to select the correct tip size for a positive pressure torch.

Standard Tip Cleaner Sizes	
Manufacturer's Code	**Corresponding Drill Size**
6	77-76
7	75-74
8	73-72-71
10	70-69-68
12	67-66-65
14	64-63-62
15	61-60
16	59-58
17	57
18	56
22	55-54
24	53-52
26	51-50-49

Figure 7-11. Tip cleaner numbers by manufacturer's code and drill size.

- Avoid allowing the flame end of the tip to come in contact with the welding work, bench, or firebricks.
- Never use the torch tip to push, lift, or strike the piece being welded.

Check Your Progress

Write your answers in the spaces provided.

1. What two functions does the welding torch perform?

   a. _____

   b. _____

2. List the five main parts of the welding torch.

   a. _____

   b. _____

   c. _____

   d. _____

   e. _____

3. What are the two types of torches commonly used for oxyfuel gas welding?

   a. _____

   b. _____

4. Acetylene pressures from _____ to _____ are used with the medium pressure torch.

5. The size of a welding tip can be determined by the _____ of the orifice.

6. What is the purpose of a flashback arrestor? _____

   _____

7. Why and how should the torch tip be regularly cleaned?

   _____

   _____

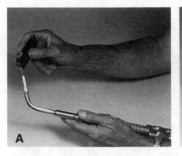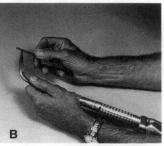

Figure 7-12. Caring for welding tips. A—Using tip cleaner to clean tip orifice. B—Reconditioning the orifice end of a welding tip.

8. Welding tips are made of _____, which transfers the heat of the flame away from the torch tip.

9. Each manufacturer's tip number indicates (circle letter):
   a. the length of the tip.
   b. the size of the flame.
   c. a given tip angle.
   d. a given drill size.

10. Which is affected by the size of the orifice in the torch tip (circle letter)?
   a. Temperature of the welding flame.
   b. Amount of heat produced by the flame.
   c. Color and length of the flame.

Instructor's Initials _____ Date _____

Things to Do

1. With the approval of your instructor, disassemble a welding torch and examine the parts. Try to determine if it is an injector- or medium pressure–type torch.

2. Prepare a large poster for the bulletin board showing the differences between the mixing chambers of the medium pressure– and injector-type welding torches. Use different colors to show how the gases enter the chamber and where they are mixed.

3. Write to welding equipment manufacturers for specifications on the styles of torches they produce. Create a collection of literature on welding torches.

4. Check torch tips in your welding area. Clean tips that are in poor condition using the correct size tip cleaner. If a tip appears to be damaged, show it to your instructor.

5. Prepare a chart that compares the manufacturer's tip number with a standard drill size for all welding tips in your shop. Carefully observe how each company uses a different numbering system to designate tip sizes.

6. Make a poster of things to do or not do to keep welding tips in good condition. Ask permission to mount the poster on a wall in the welding area as a reminder to all welders.

Unit 8

ASSEMBLY OF WELDING EQUIPMENT

Oxyfuel Gas Welding Equipment

Safe, high-quality work starts with an understanding of welding equipment and setup procedures. Once a complete oxyfuel gas welding outfit is correctly assembled, regular maintenance and inspection is essential. Equipment will operate at its maximum potential and produce quality welds for many hours if it is properly cared for.

A modern gas welding outfit consists of several components, **Figure 8-1:**

- Cylinder of oxygen gas.
- Cylinder of acetylene gas.
- Oxygen regulator.
- Acetylene regulator.
- Two lengths of colored hose, green or black for oxygen and red for acetylene.
- Torch body.
- Two flashback arrestors installed between the hoses and the body.
- Assorted tips.

Figure 8-2 shows the hoses, flashback arrestors, torch body, and torch tip correctly assembled.

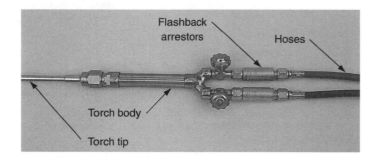

Figure 8-2. Correct assembly of the hoses, flashback arrestors, torch body, and welding tip. (Victor Equipment Co.)

Pressure Regulators

Pressure regulators reduce the high-pressure gases in the cylinders to low pressures that are usable for welding operations. See **Figures 8-3** and **8-4.** You will recall a fully charged oxygen cylinder contains 2200 psig (15 170 kPa) of pressure, and an acetylene cylinder contains 250 psig (1720 kPa) of pressure.

When the oxygen and acetylene cylinder valves are opened, pressures register on each respective *cylinder pressure gauge.*

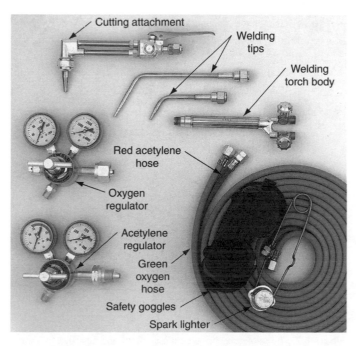

Figure 8-1. Essential oxyfuel gas welding equipment. (Uniweld Products, Inc.)

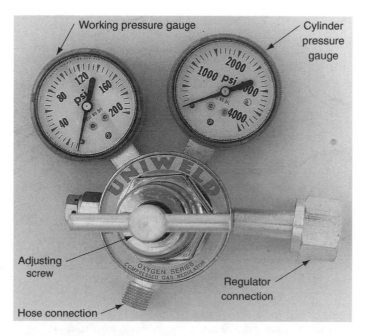

Figure 8-3. Typical oxygen regulator with gauges. (Uniweld Products, Inc.)

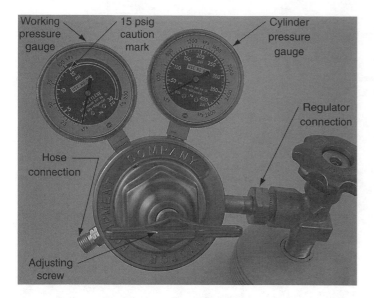

Figure 8-4. Acetylene regulator attached to an acetylene cylinder. (Victor Equipment Co.)

The readings indicate the amount of pressure and gas in the cylinders. The adjusting screw is turned clockwise until the desired working pressure is indicated on each *working pressure gauge*. Working pressure is the pressure that flows to the torch tip. An acetylene pressure gauge has a red background or red markings starting at 15 psig (103 kPa), a visual reminder not to exceed this amount of pressure.

Setup Procedures

To properly set up an oxyfuel gas welding outfit, follow these procedures:

1. Securely fasten both cylinders with chains or straps in the vertical position to a portable welding cart or a sturdy fixture. Remove the cylinder safety caps.

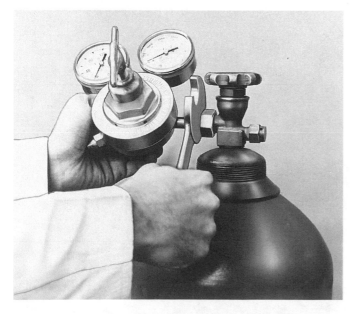

Figure 8-5. Tightening a regulator connection to the cylinder valve with a regulator wrench. (Victor Equipment Co.)

Figure 8-6. Connecting a hose to the regulator.

2. Crack (quickly open and close) each cylinder valve to blow out any dirt that could be carried to the regulator. Be sure the stream of gas is not directed toward another person or an open flame.

3. Connect each pressure regulator to the cylinder nozzle. The connecting threads on an oxygen regulator are right-handed, while the connecting threads on a fuel gas regulator are left-handed. Pressure regulators should be threaded onto the cylinder nozzles by hand and tightened with a *regulator wrench,* **Figure 8-5.**

4. Attach the oxygen and acetylene hoses to their respective regulators, **Figure 8-6.** Acetylene hoses are red; oxygen hoses are green or black, making them easy to distinguish. The connections are also different to prevent the wrong connection. Oxygen hose connections, like those of the regulator, have right-handed threads, while acetylene hose connections have left-handed threads. Acetylene hose connection nuts have a groove around the center, **Figure 8-7.**

5. Attach the torch to the other end of the hoses by hand tightening. (The torch has right- and left-handed threads like the regulators and hoses.) Carefully tighten the connections using proper hose and regulator wrenches, which seal the connections without damaging the equipment.

6. Select the desired welding tip and attach it to the torch, **Figure 8-8.**

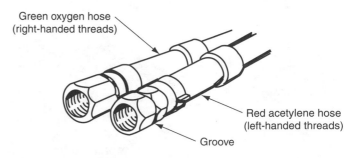

Figure 8-7. The acetylene hose connection nut has a groove to help distinguish it from the oxygen hose connection nut. (Linde Co. Div. Union Carbide Corp.)

Figure 8-8. Fastening a tip assembly to the torch body.

Before using an oxyfuel gas welding outfit, visually inspect the cylinders, pressure regulators, hoses, fittings, and torch for possible damage. Worn or faulty equipment can cause accidents.

The hoses and regulator connections can be safely checked for leaks using a nonpetroleum soap solution. Wipe the hoses and connections with the soapy liquid and look for bubbling. The location of the bubbles indicates a gas leak. *Never use a flame to check for leaks!*

Once the equipment is thoroughly checked, the torch is ready to be lit. (Lighting the torch is discussed in Unit 11.) **Figure 8-9** shows an entire oxyfuel gas welding outfit ready for use.

Check Your Progress

Write your answers in the spaces provided.

1. List the eight pieces of equipment assembled to make a modern welding outfit.

   a. _____

   b. _____

   c. _____

   d. _____

   e. _____

   f. _____

   g. _____

   h. _____

2. Explain the purpose of a pressure regulator. _____
   _____
   _____

3. _____ pressure is the pressure of the gas that flows to the torch tip.

4. In which direction is the regulator adjusting screw turned to increase the pressure of the gas delivered to the torch?
   _____

5. Why is it a good practice to crack each cylinder valve before connecting the regulator? _____
   _____
   _____

Figure 8-9. A complete welding outfit ready for use. (Linde Co. Div. Union Carbide Corp.)

6. Acetylene should never be used at pressures above _____.

7. Pressure regulators should be threaded onto the cylinder nozzles by hand and tightened with a(n) _____.

8. It is easy to tell which hose is used for acetylene gas. The color of the hose is _____.
   a. black
   b. blue
   c. red
   d. green

9. Oxygen hose connections have _____ threads, and fuel gas hose connections have _____ threads.

10. Describe a safe way to check for leaks in an oxyfuel gas setup. _____
    _____
    _____

Instructor's Initials _____ Date _____

Things to Do

1. On a separate sheet of paper, sketch the pieces of equipment that make up a welding outfit. Use arrows and names to indicate each piece.

2. With your instructor, go through the correct steps in setting up an oxyfuel gas welding outfit. Practice these steps until you become very familiar with the process.

3. Check the welding outfits in your shop to see if all cylinders are securely fastened in a safe, vertical position. If any cylinders do not appear safe, tell your instructor.

4. Carefully examine a set of oxygen and fuel gas regulators to better understand how cylinder pressures are recorded and working pressures are set by the adjusting screw.

METAL IDENTIFICATION

Importance of Identification

Identifying metals for welding purposes requires a great deal of practice. Some metals, such as copper or aluminum, are recognizable after careful examination. Others are not so simple to identify. Because there are so many types of metals, exact identification can be difficult. Sometimes a metal must be tested in a laboratory to determine its actual identity.

Welders must be able to identify the most common metals to select the correct welding or brazing rod for a particular joint. With experience, welders learn to gather information about metals for quick identification.

Ferrous or Nonferrous

Metals are classified as either ferrous or nonferrous. *Ferrous metals* contain iron or iron with additional elements. Ferrous metals include mild steel, low- to high-carbon steel, tool steel, stainless steel, and cast irons. Most industrial welding is done on ferrous metals.

Iron is the primary element in ferrous metals. Carbon is added to iron to make steel. Additional elements, called *alloying elements,* are added to create the various types of steels. An *alloy* is a mixture of two or more metals. Stainless steel, for example, is an alloy of iron along with carbon, nickel, chromium, and small amounts of other alloying elements.

Nonferrous metals do not contain significant amounts of iron but may contain small amounts of iron as an alloying element. Nonferrous metals include aluminum, copper, tin, lead, and zirconium. Other nonferrous metals, such as brass (copper and zinc) and bronze (copper and tin), are alloys. Valuable nonferrous metals, such as nickel, chromium and vanadium, are seldom welded. Instead, they are used as alloying elements in ferrous and nonferrous metals to produce alloys for specific applications.

Identification Procedure

Welders cannot be expected to do scientific testing for metal identification. However, certain methods can help them identify the metals they are expected to weld. The best way to identify a metal is to obtain it from a known source. Then, the type of metal is known. Label or tag the metals to identify them.

Steel manufacturers color code the ends of steel bars and rods. When using new steel from a storage area, check the color against the manufacturer's specifications to determine grade.

When working in metal fabrication, a drawing or print indicating the metals to be used is often available. Metal type is written in the title block or bill of materials, **Figure 9-1.**

Identification Tests

If the metal is not known, a number of tests can be performed. Identification tests include:
- Color test.
- Density test.
- Magnetic test.
- Spark test.
- Hardness test.
- Oxyacetylene torch test.

The common nonferrous metals used in welding (brass, bronze, and aluminum) can be identified by a simple *color test.* Brass is usually bright yellow. Bronze is copper-colored. Aluminum looks much like silver. Unlike nonferrous metals, ferrous metals look alike and cannot be easily distinguished by color.

A *density test* identifies a metal by determining its relative weight. For example, aluminum, lead, steel, and stainless steel are similar in appearance but not weight. Aluminum is the lightest metal; lead is the heaviest. Stainless steel is only slightly heavier than steel.

UNLESS OTHERWISE SPECIFIED DIMENSIONS ARE IN INCHES TOLERANCES ON FRACTIONS ± 1/64 DECIMALS ± 0.010 ANGLES ± 1°	**BAIRD PRODUCTS**		
	TITLE WELDING TABLE		
	SCALE FULL	CHK'D AMR	DATE 10-3-04
MATERIAL MILD STEEL	DRAWN BY RJB	DRAWING NO. B-205588	

Figure 9-1. Metal specification shown on the title block of a typical drawing.

Figure 9-2. The welder is performing a spark test to determine the grade of a piece of steel.

The **magnetic test** can quickly separate most ferrous and nonferrous alloys. Metals containing iron will be attracted by the magnet; nonferrous metals will not. One exception is the 300-series of stainless steels. Although this family of stainless steels is ferrous, it is not magnetic.

Color, density, and magnetic tests are useful for quickly identifying a metal. When more complex testing is required for precise identification, the spark, hardness, and oxyacetylene torch tests are used.

A **spark test** can be used to accurately identify different grades of steel and cast iron. To perform a spark test, firmly press a piece of steel or cast iron against a rotating grinding wheel, **Figure 9-2.** Observe the color, length, and shape of the stream of sparks. The different alloying elements give visual clues as to the identity of each type of metal. See **Figure 9-3.** A spark test should be performed in a darkened area so the sparks can be clearly seen. Always wear safety glasses and gloves when working with grinding equipment.

Certain grades of steel can be identified by a **hardness test.** The welder determines how easy or difficult it is to cut the steel with a file. Identification is based on the following results:

- Mild steel is fairly soft and cuts easily.
- Low- and medium-carbon steels can be cut, but with some difficulty.
- Alloy steels require considerable pressure to cut.
- Tool steels can be cut with a great deal of pressure.
- Hardened alloy or tool steels cannot be cut.

An **oxyacetylene torch test** can be performed on ferrous metals to determine if they are weldable. A small molten pool is made in the metal. If the steel is weldable, the pool will be fluid and emit few sparks. If the steel is not weldable, the pool will be sluggish and emit many sparks.

Check Your Progress

Write your answers in the spaces provided.

1. Why is it important for the welder to be able to identify the most common metals? _____

_____

_____

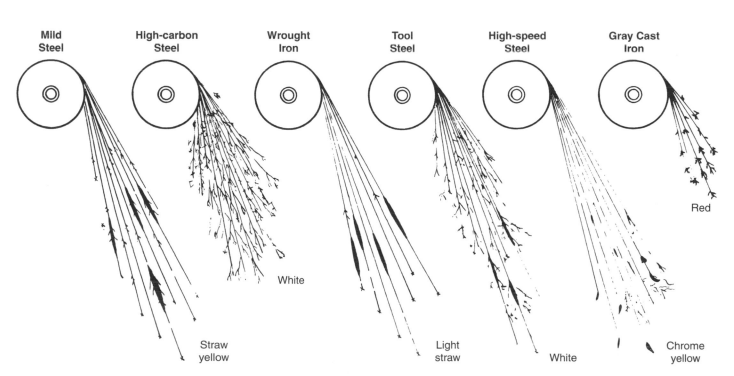

Figure 9-3. The color, length, and shape of the stream of sparks generated by a spark test can help the welder determine the identity of irons and steels.

2. Metals are classified as either _____ or
_____.

3. What tests are useful for quickly identifying a metal, but
are not precise? _____

_____

4. Different grades of steel and cast iron can be accurately
identified using a(n) _____ test.

5. List the five grades of steel that can be tested for hardness
by cutting with a file.

a. _____

b. _____

c. _____

d. _____

e. _____

Instructor's Initials _____ Date _____

Things to Do

1. Gather various types of metal from around the shop. Using
the tests described in this unit, try to identify each piece of
metal. Make a record of the testing methods used.

2. Ask your instructor for two different grades of steel. Place
each one against a grinding wheel. Compare the sparks
from each piece to those in Figure 9-3.

3. Use the oxyacetylene torch test on various pieces of steel
to determine if the steel is weldable. Describe your results.

Unit 10

WELDING RODS AND FLUXES

Filler Metal

Many weld joints require the addition of metal to fill or complete a weld joint. Welding rod (also called *filler metal*) is the metal added to a weld, braze, or solder joint. *Fluxes* are used to make welding possible on some metals. Fluxes clean the metal and protect it during welding, brazing, or soldering.

Filler metals used for oxyfuel gas welding are either ferrous or nonferrous, just as base metals are. The ferrous group of filler metals includes steel and cast iron. The nonferrous group includes aluminum, copper, brass, and bronze.

Classification and Size

The American Welding Society (AWS) classifies welding rods used for oxyfuel gas welding, **Figure 10-1.** Each type of welding rod is designated by RG. The letter R stands for welding rod; the letter G stands for gas welding. The rods are copper-coated to prevent rusting during storage. Because of the copper coating, oxyfuel gas welding rods are not to be used as filler rods for other welding processes.

Filler rods come in a variety of diameters, **Figure 10-2.** Generally, they are straight, 36″ (91 cm) long, and packaged in 50 lb. (23 kg) cartons or 5 lb. and 10 lb. (2 kg and 5 kg) tubes. Aluminum welding rods also come in coils, **Figure 10-3.** Suggested rod sizes for various metal thicknesses are shown in **Figure 10-4.**

Rod Selection

Factors to consider when selecting a welding rod are:
- Type of metal being welded.
- Thickness of the metal being welded.
- Type of joint being made.
- Size of the weld.
- Strength of the required weld.

The type of rod to be used depends upon the base metal being joined. Some common welding rods are:
- Low-carbon steel.
- High-strength steel.
- Stainless steel.
- Cast iron.
- Aluminum.
- Brass alloy.

Low-carbon steel welding rod is used to weld mild steel for most general purposes. High-strength steel rod is used to weld mild steel to produce stronger joints. For example, steel tanks or pipes used under high pressure are welded with high-strength steel rod. It is also used for welding high-carbon steel.

AWS Classification	Minimum Tensile Strength	Application
RG 45	45,000 psi (310 kPa)	General usage on low-carbon steels
RG 60	60,000 psi (414 kPa)	Mild steel with 50,000 psi to 65,000 psi (345 kPa to 448 kPa) tensile strengths
RG 65	65,000 psi (448 kPa)	Mild and low-alloy steels with 65,000 psi to 75,000 psi (448 kPa to 517 kPa) tensile strengths

Figure 10-1. AWS designations for oxyfuel gas steel welding rods.

Cast iron rod should be selected according to manufacturers' specifications. The designations for cast iron welding rods are RCI, RCI-A and RCI-B. The letters CI stand for cast iron. The designations -A and -B are different alloys of cast iron. Check manufacturers' specifications for the required flux.

Aluminum welding rod is used for welding and brazing aluminum alloys and castings. The most common aluminum rods are designated by the numbers 1100, 4043, and 5356. A flux is required when oxyfuel gas welding aluminum.

Diameter	Actual size	Steel	Brass (Bronze)	Aluminum	Cast iron (square)
1/16″	●	31	29	91	—
3/32″	●	14	13	41	—
1/8″	●	8	7	23	—
5/32″	●	5	—	—	—
3/16″	●	3 1/2	3	9	5 1/2
1/4″	●	2	2	6	2 1/4
5/16″	●	1 1/3	—	—	1/2
3/8″	●	1	1	—	1/4

Figure 10-2. Available sizes and approximate number of welding rods per pound. (Airco Welding Products, Div. Air Reduction Co.)

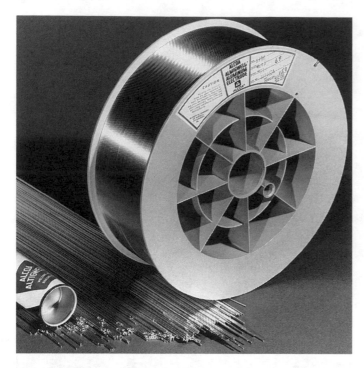

Figure 10-3. Aluminum coil and welding rods. (ALCOA)

Metal Thickness	Welding Rod Diameter
1/16″ (1.6 mm)	1/16″ or 3/32″ (1.6 mm or 2.4 mm)
1/8″ (3.2 mm)	3/32″ or 1/8″ (2.4 mm or 3.2 mm)
1/4″ (6.3 mm)	1/8″ or 3/16″ (3.1 mm or 4.8 mm)
3/8″ (9.5 mm)	3/16″ or 1/4″ (4.8 mm or 6.3 mm)

Figure 10-4. Guide to rod selection for various metal thicknesses.

Brass alloy rod (often referred to as bronze rod) is used for welding copper alloys and brazing other metals. It is also used to braze many grades of steel and cast iron. A flux is required either separately or as a coated rod.

Welding Fluxes

When most metals are heated to the molten (liquid) state, they absorb oxygen from the air and oxidize. The oxides mix with the molten metal and, when solidified, weaken the joint. Fluxes remove oxides and other impurities from the joint by floating them to the surface of the molten weld pool.

Different types of fluxes are produced for use with stainless steel, cast iron, aluminum, brass, and bronze. The fluxes are available in liquid, powder, or paste form. Many filler rods come with a flux coating, eliminating the need to continuously dip the rod in a flux while welding. A flux is not required when welding steel with steel filler rod. Fluxes are used when brazing steel.

Check Your Progress

Write your answers in the spaces provided.

1. Filler metals are grouped into the same two classifications as base metals: _____ and _____.

2. Why do oxyfuel gas welding rods have a copper coating? _____

3. What do the letters RG stand for in the filler rod designation RG 45? _____
_____

4. What diameter of welding rod is recommended for welding 1/8″ metal? _____
_____

5. Five factors to consider when selecting a welding rod are:
   a. _____
   b. _____
   c. _____
   d. _____
   e. _____

6. List six common filler rods used in welding and brazing.
   a. _____
   b. _____
   c. _____
   d. _____
   e. _____
   f. _____

7. Is a flux required to weld each of the following processes? (Circle *Yes* or *No.*)
   a. Welding mild steel.   *Yes*   *No*
   b. Welding cast iron.   *Yes*   *No*
   c. Welding aluminum.   *Yes*   *No*
   d. Brazing steel.   *Yes*   *No*

8. What type of welding rod is used to weld mild steel to produce stronger joints? _____
_____

9. What happens to most metals when heated to the molten state? _____
_____
_____

10. The purpose of flux is to remove _____ and other _____ from the joint and float them to the surface of the weld pool.

Instructor's Initials _____ Date _____

Things to Do

1. With your instructor, compare the various types and diameters of welding rod used in your shop. Write a report on the proper welding rods to use with various welding processes.

2. Contact a local welding supply dealer for information on how welding rod is packaged. Also obtain a current price list for welding rod.

This welder is using a cutting torch to cut a steel plate. She is wearing proper protective clothing. (Printed with Permission of the American Welding Society)

Unit 11

THE WELDING FLAME

Producing a Flame

The welding flame is the welder's most important tool. The welding equipment described in previous units serves to produce the flame. To carry out each welding task with the greatest efficiency, the proper size, shape, and color of the welding flame must be monitored and maintained.

Many types of fuel gases are used to produce flames for various welding processes. Besides acetylene, other gases such as propylene, MAPP (methylacetylene propadiene), propane, methane (natural gas), and hydrogen are used. The flames of these various oxyfuel gas combinations have been tested to determine the physical properties of the welding flame, **Figure 11-1.** Of all the fuel gases that can be combined with oxygen to produce a welding flame, acetylene is the most widely used.

A chief advantage of the oxyacetylene flame is its high operating temperature. The proper mixture of oxygen and acetylene burns at 5589°F (3087°C), which makes it the hottest oxyfuel gas flame. An oxyacetylene flame can easily melt all common metals.

Types of Flames

Three types of flames can be produced with the oxyacetylene torch: carburizing, neutral, and oxidizing. See **Figure 11-2.** The operator of the oxyacetylene welding outfit must master the adjustment of the oxygen and acetylene hand valves on the welding torch to produce the desired flame. The operator also must be able to recognize each type of flame by its size, shape, and color and to understand the effect of the oxyacetylene flame on the metals to be joined.

Carburizing Flame

The *carburizing flame* contains more acetylene than oxygen. This flame is easy to recognize by the outer flame, which has a slim bluish taper with an orange tip. Inside this outer flame is a shorter, light blue flame called the acetylene feather. Its length depends upon the amount of excess acetylene allowed to flow from the tip. The carburizing flame is sometimes described as an "oxygen-starved" flame because little oxygen is present. Within the acetylene feather is the inner cone. It is a short, brilliant white flame with a somewhat ragged point.

Figure 11-1. Flame testing is done to determine the physical characteristics of various oxyfuel gas flames. (American Torch Tip Co.)

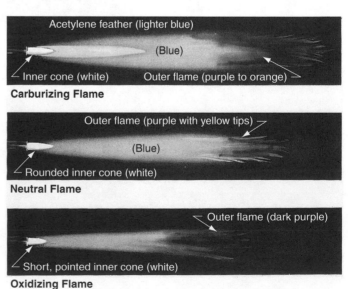

Figure 11-2. Characteristics of the three welding flames. (Smith Welding Equipment Co.)

The carburizing flame has a lower temperature than the neutral flame. The carburizing flame is seldom used to weld steel because the high carbon content in the flame is absorbed by the hot steel. This unwanted addition of carbon changes the structure of the steel, giving it the brittleness and hardness of high-carbon steel when cooled. Carburizing flames are sometimes used for silver brazing, braze welding, and soldering.

Neutral Flame

The *neutral flame* is produced by a proper balance of oxygen and acetylene. It is approximately 5589°F (3087°C) and is the most widely used of the three types of flames. The flame is called neutral because it requires 1 part acetylene to 1 part oxygen from the cylinders. However, to complete the combustion, 2 1/2 parts oxygen to 1 part acetylene are required. The surrounding air supplies 1 1/2 parts of the needed oxygen.

As the flame is adjusted from a carburizing flame to a neutral flame by adding more oxygen, the light blue acetylene feather disappears. At the same time, the brilliant inner cone becomes slightly shorter and more sharply defined. The outer bluish flame decreases slightly in size and loses most of the orange tip. At this point the flame becomes neutral.

The neutral flame provides several advantages for fusion welding. When the flame is directed upon a steel surface, the metal slowly melts. See **Figure 11-3.** The molten metal does not boil and few sparks appear. The neutral flame protects the molten steel from oxidation (taking on more oxygen), and the weld joint is not burned. No carbon is added or taken away from the steel since the neutral flame burns up all the carbon from the acetylene gas. Extremely strong welds are produced in this manner.

Oxidizing Flame

The *oxidizing flame* contains more oxygen than acetylene. The shape of the oxidizing flame is similar to the shape of the neutral flame. However, the inner cone of the oxidizing flame is more pointed, shorter, and purple. The outer flame is slightly shorter, flared at the end, and almost completely blue. The oxidizing flame can also be distinguished from the neutral flame by a loud hissing sound caused by the excess rush of oxygen.

The oxidizing flame has very little practical use in welding. Its temperature may reach 6300°F (3480°C), but the

Figure 11-3. A neutral oxyacetylene flame is being used to heat steel bar.

oxides that form when the excess oxygen combines with the molten metal cause problems with the weld. Such problems include brittleness, low strength, and hardening around the weld area.

Lighting the Torch

Recall from Units 7 and 8 that torch valves (also called finger-tip valves or hand valves) are located on the torch body. Cylinder valves are located on the top of each gas cylinder. Regulator adjusting screws are located on the front of each regulator. Knowing the names and locations of these valves and adjusting mechanisms is necessary before attempting to light a torch.

Follow these procedures when lighting a welding torch and adjusting the flame:

1. Check that the oxygen and acetylene torch valves are closed by turning then in a clockwise direction.

2. Check that the oxygen and acetylene regulator adjusting screws are turned counterclockwise and move freely. Do not turn the adjusting screws too far, or they will come off.

3. Slowly open the acetylene cylinder valve 1/4 to 1/2 turn in a counterclockwise direction. Use the proper valve wrench, and leave it on the valve so you can quickly shut off the gas in an emergency. **Warning:** Do not stand directly in front of the regulator face when opening the cylinder valve. The high-pressure gas could burst the regulator and cause injury.

4. Slowly open the oxygen cylinder valve as far as it will go.

5. Set the correct working pressure on the acetylene regulator gauge. Do this by opening the acetylene torch valve and turning the acetylene regulator adjusting screw until the desired pressure is reached. Close the acetylene torch valve.

6. Repeat Step 5 to set the oxygen working pressure.

7. Open the acetylene torch valve 3/4 turn.

8. Using a standard spark lighter, light the acetylene gas, **Figure 11-4.** Adjust the flame until it burns turbulently approximately 3/4″ (19 mm) from the torch tip. Then, adjust the flame so no sooty smoke is released. **Warning:** Always use a spark lighter to light the torch. Never use a match or butane lighter. Your hand could be severely injured when the gas fumes are ignited.

9. Slowly open the oxygen torch valve, **Figure 11-5.** Adjust the oxygen until a small pointed cone appears at the torch tip. This is a neutral flame. Minor adjustments in the torch valves and regulator pressures may be necessary to obtain a neutral flame.

Shutting Off the Torch

Whenever the welding torch is not in your hand, it should be shut off. Turn off the acetylene torch valve first, then the oxygen torch valve. If you are leaving the welding area or are

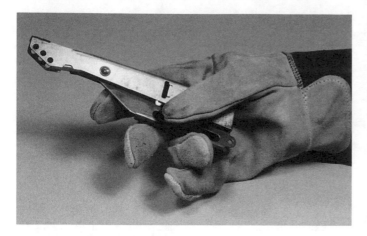

Figure 11-4. Two styles of spark lighters that use flint to produce the spark.

finished for the day, completely shut down the equipment by following these steps:

1. Close the acetylene torch valve on the torch body to extinguish the flame. Then, close the oxygen torch valve so gas is not being released from the torch tip.

2. Close both the acetylene and oxygen cylinder valves.

3. Open both torch valves on the torch body to release any gas still in the hoses.

4. Watch the acetylene and oxygen regulators until you see that the high- and low-pressure gauges on both regulators read zero.

5. Once the gauges read zero, close the acetylene and oxygen torch valves and put the torch away.

6. Turn the adjusting screws on both regulators counterclockwise until they turn freely.

Check Your Progress

Write your answers in the spaces provided.

1. Describe the skills that must be mastered by the operator of the oxyacetylene welding outfit.

   a. _____

   _____

   b. _____

   _____

   c. _____

   _____

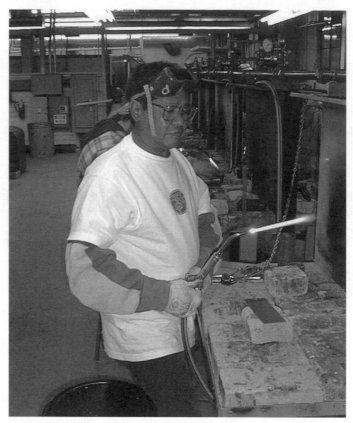

Figure 11-5. Adjusting the oxyacetylene flame to a neutral flame.

2. Describe the appearance of a carburizing flame. _____

   _____

   _____

   _____

3. For complete combustion, the neutral flame requires _____ part(s) oxygen from the surrounding air.

4. The temperature of the neutral flame is _____.

5. List five advantages of using the neutral flame for fusion welding.

   a. _____

   b. _____

   c. _____

   d. _____

   e. _____

6. Why is the neutral flame called "neutral"? _____

   _____

7. Describe the appearance of a neutral flame. _____

   _____

   _____

   _____

8. Three problems caused by using an oxidizing flame for fusion welding are:

   a. _____

   b. _____

   c. _____

9. Put the following nine procedures for lighting the oxyfuel torch in proper order. Write the number of the step in each blank.

   ____ Open the acetylene torch valve 3/4 turn.

   ____ Slowly open the oxygen cylinder valve as far as it will go.

   ____ Slowly open the acetylene cylinder valve 1/4 to 1/2 turn.

   ____ Set the correct working pressure on the oxygen regulator gauge.

   ____ Check that the oxygen and acetylene torch valves are closed.

   ____ Slowly open the oxygen torch valve and adjust to a neutral flame.

   ____ Set the correct working pressure on the acetylene regulator gauge.

   ____ Using a standard spark lighter, light the acetylene gas.

   ____ Check that the regulator adjusting screws move freely.

10. *True or false.* The first step in shutting off the torch is to close the acetylene torch valve (before closing the oxygen torch valve). _____

Instructor's Initials _____ Date _____

Things to Do

1. Light the torch and adjust it to an oxidizing flame. Be sure to wear goggles and protective clothing. Direct the cone close to a piece of scrap metal until the surface of the metal begins to melt. Carefully observe the molten (melted) area. In the space below, sketch what you see.

2. Adjust the torch to a carburizing flame and repeat the previous activity. In the space below, sketch what you see.

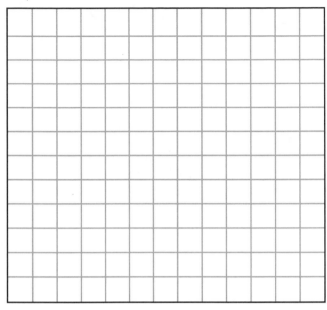

3. Adjust the torch to a neutral flame and repeat the previous activity. In the space below, sketch what you see.

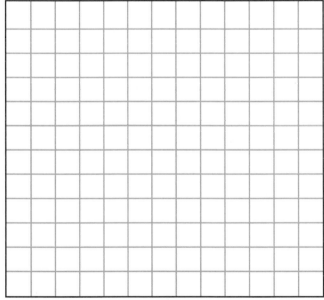

4. Practice lighting the torch and adjusting it to produce a neutral flame. Use a variety of tip sizes.

Unit 12

BACKHAND AND FOREHAND WELDING

Manipulating the Torch

Two techniques are used in manipulating (handling) the torch and welding filler metal in oxyfuel gas welding. They are known as backhand and forehand welding. Forehand welding is the easier and more commonly used method. However, each has certain advantages for different metal thicknesses.

Forehand Method

In the *forehand method,* welding moves from right to left (for a right-handed person). The welding filler metal is moved ahead of the torch tip and away from the completed portion of the weld, **Figure 12-1.** The welding flame is pointed in the direction of travel and at a downward angle. The flame melts the edges of the joint and preheats the welding filler metal. *Welding rod* (also called filler metal or filler rod) comes in long, straight lengths for use in oxyfuel gas welding.

The torch tip is moved in a circular, semicircular, or weaving path to evenly distribute the heat and molten metal, **Figure 12-2.** Each of these patterns is equally effective in evenly distributing the heat. The key is to maintain a consistent, steady pattern for the length of the weld. Forehand welding gives good fusion on a joint with or without using filler metal. It is ideal for welding thin sheet steel.

On thick steel, a large weld pool is formed when the walls of the deep joint are melted. The pool can be difficult to control using the forehand technique because it requires an included

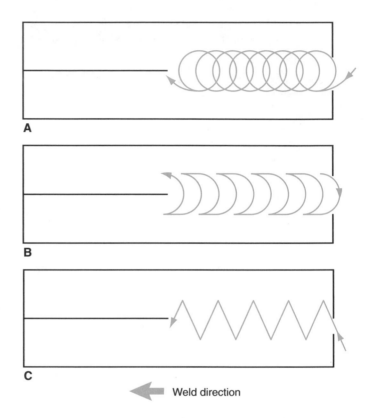

A

B

C

Weld direction

Figure 12-2. Motion of the torch tip in forehand welding. A—Circular pattern. B—Semicircular pattern. C—Weave pattern.

joint angle of approximately 90° to obtain complete fusion. An *included joint angle* is formed by the angles of the edges of the two pieces of metal being welded. It is the total angle, measured in degrees, of the joint to be filled.

Backhand Method

The *backhand method* moves from left to right (for a right-handed person). The torch tip moves ahead of the filler metal in the direction the weld is being made, **Figure 12-3.** The filler metal is placed between the torch flame and the completed weld. The flame is pointed back at the filler metal and the molten weld pool.

In backhand welding, the movements of the welding rod and torch flame differ from the movements used in forehand welding. The welding rod may be moved back and forth through the weld pool, or it may be moved in a circular motion. The torch flame is held fairly still in the middle of the groove. Just enough motion is used to keep the weld pool straight. A slow, even movement of the flame in the direction of welding produces good results.

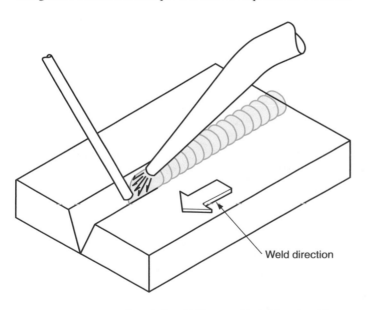

Weld direction

Figure 12-1. Position of torch tip and filler metal in forehand welding.

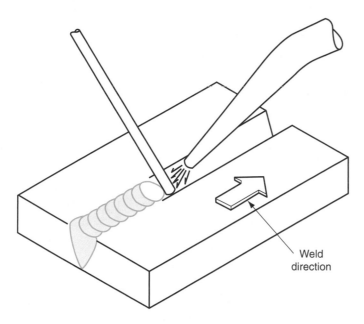

Figure 12-3. Position of the torch tip and filler metal in backhand welding.

As you practice making a variety of welded joints, the following advantages of backhand welding will become apparent to you:

- Smaller included joint angles on a V-groove can be used in joint preparation. Complete fusion can be achieved with a 60° V-groove, rather than a 90° V-groove as used in the forehand technique.
- A smaller weld pool can be used. The smaller pool is easier to control and good surface fusion can still be obtained.
- Less welding filler metal is used because of the narrow V; therefore, large welds can be completed in less time.
- Less oxygen and acetylene are used.
- Less oxidation results since the flame protects the weld pool as it is directed back toward the weld.

Holding the Torch

In both the forehand and backhand techniques, the torch can be held in one of two ways. One option is to grasp the torch like a hammer, using an overhand grip. The other option is to grasp it like a pencil, using a delicate underhand grip by the thumb, index finger, and middle finger. Place the hoses over your arm or shoulder so only the weight of the torch body is being held and supported during welding. Doing so will reduce fatigue and increase steadiness so a consistent flame pattern can be applied.

Check Your Progress

Write your answers in the spaces provided.

1. Three torch tip manipulations used in forehand welding are:

   a. _____

   b. _____

   c. _____

2. The direction of the weld in the backhand method progresses from _____ to _____ for right-handed welders.

3. Explain why a large weld pool is created during forehand welding of thick steel. _____

   _____

   _____

   _____

4. List one advantage and one disadvantage of the forehand welding technique. _____

   _____

   _____

   _____

5. Circle the letter(s) of the statements that apply to backhand welding:

   a. Large welds can be completed in less time.

   b. More filler metal is used to fill the joint.

   c. Smaller Vs can be used in joint preparation, usually a 60° included angle.

   d. Torch tip motion should make large circles or semicircles.

Instructor's Initials _____ Date _____

Things to Do

1. Prepare a sketch showing the backhand method of making a fillet weld on a T-joint. Show the completed part of the weld, weld pool, filler metal, and torch tip in the correct position.

2. Sketch the position of the torch tip and filler metal for forehand welding on the joints below. Show the direction of welding.

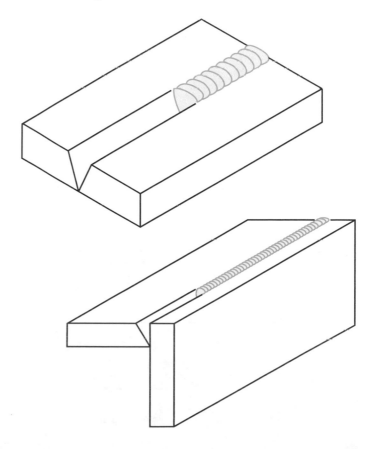

Unit 13

CREATING A CONTINUOUS WELD POOL

Basic Operations

Creating a **continuous weld pool** refers to the process of moving a molten pool of metal across the surface of a piece of metal with or without the use of welding rod. The **weld pool** refers to the small molten area formed below the tip of the oxyfuel gas welding flame. Creating a continuous weld pool provides the opportunity to practice the following basic operations:

- Adjusting the torch valves to produce a neutral flame.
- Manipulating the torch body and directing the flame.
- Controlling the size and direction of the weld pool.
- Discovering the proper travel speed, work angle, and travel angle needed to produce a weld bead with uniform width and pattern. A **weld bead** results from running a continuous weld pool and allowing it to solidify (become solid).

Preparing the Equipment and Material

Before beginning the process of creating a continuous weld pool, the following preparations must be made:

1. Visually inspect the welding outfit to be sure it is in good condition.

2. Clear the welding area of any flammable materials.

3. Put on safety goggles, gloves, and protective clothing before beginning to weld.

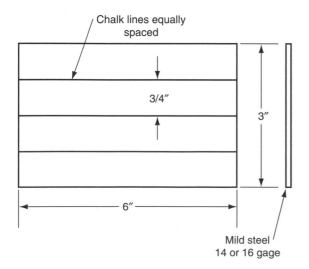

Chalk lines equally spaced

3/4″

3″

6″

Mild steel 14 or 16 gage

Figure 13-1. Specifications for weld pool practice material.

4. Obtain a piece of 14 gage or 16 gage (1.8 mm or 1.5 mm) mild steel measuring 3″ × 6″ (76 mm × 152 mm).

5. Clean the surface of the metal with a stainless steel brush, steel wool, or abrasive paper. Remove all sharp edges and burrs with a file or a grinder.

6. Use a soapstone and a ruler to draw three equally spaced lines on your practice piece, **Figure 13-1.**

7. Select the correct torch tip size for the metal thickness. Remember, tip sizes are numbered according to the manufacturer's specifications.

8. Place the sheet piece on top of fire bricks on the welding table in a position comfortable for forehand welding.

Procedures for Running a Continuous Bead

Follow these procedures for making a continuous weld pool:

1. Set both the oxygen and acetylene pressures to 6 psig (41 kPa) on the regulators. Light the torch and adjust the flame to neutral, **Figure 13-2.**

2. Hold the torch in a comfortable, balanced position. Point the flame in the direction of the weld at a 90° work angle and a 35° to 45° travel angle. See **Figure 13-3.** (In this text, torch angle is stated by giving two angles: the **work angle** and the **travel angle,** a method designated by the American Welding Society.)

3. Begin the weld pool at the right-hand edge of the metal (for a right-handed welder). Lower the torch until the inner flame is approximately 1/8″ (3.2 mm) from the metal surface.

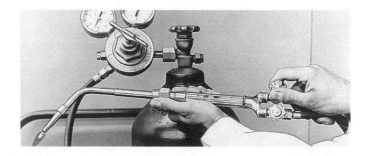

Figure 13-2. Adjusting the torch to a neutral flame. (Victor Equipment Co.)

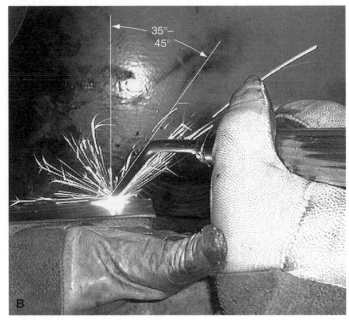

Figure 13-3. Views of continuous weld pools. A—End view with torch tip held at a 90° work angle. B—Side view with torch tip held at a 35° to 45° travel angle.

4. Keep the flame moving in a small circle until a pool of molten metal begins to form. When the pool is approximately 1/4″ (6.4 mm) in diameter, begin moving it across the sheet by slowly moving the torch in overlapping circles. Try to keep the center of the pool on the line you have drawn.

5. Slowly work the weld pool along the surface so the metal in front of the pool preheats evenly. This will produce a weld with an even, arc-shaped ripple.

6. Stop the weld approximately 1/2″ from the edge of the steel plate to prevent the edge from melting. As you weld, keep the diameter of the pool the same width and the weld as straight as possible.

7. Do a visual inspection of the first pass. Check if the width and ripple pattern are uniform for the entire length of the weld. Check the straightness of the weld bead.

8. Practice making continuous weld pools on the other two lines. Inspect each pass and try not to create the same defects you made in previous passes. Continue practicing on more pieces of metal until you can consistently and uniformly produce continuous weld pools according to proper standards.

9. Allow the welds to cool slowly, or dip them in a water tank, whichever method your instructor requires. Warning: Always pick up practice welds with pliers. The metal may be hot enough to cause serious burns. If you are not sure if a piece of metal is hot, use pliers!

Figure 13-4 shows a practice piece on which a continuous weld pool was produced. Scale (oxide) normally forms on the surrounding surface and can be removed with a wire brush. Removing the scale is not necessary when only practicing.

Finishing the Task

When you are finished welding, end the task by following these simple safety procedures:

1. Turn off the torch and hang it in the proper place at the weld bench.

2. If you are leaving the area, turn off the oxygen and acetylene cylinders. Bleed the hoses and regulators until all pressure is released. (Review the procedure in Unit 11 for shutting off the torch.)

3. Clean your welding area and put away all tools.

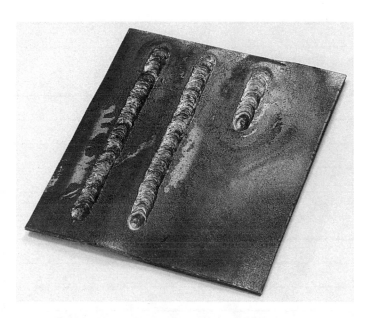

Figure 13-4. Partially completed exercise in carrying a weld pool.

Check Your Progress

Write your answers in the spaces provided.

1. Creating a continuous weld pool will give you an opportunity to practice which basic welding operations?

   a. _____

   b. _____

   c. _____

2. Always be sure to wear _____, _____, and _____ when cutting and welding.

3. To create a continuous weld pool, point the flame in the direction of welding at a work angle of _____ and a travel angle of _____.

4. When practicing weld pools, the inner flame should be approximately _____ from the metal surface.

5. When you are not using the welding torch, *(circle letter)*:
   a. hang it over the welding cart.
   b. hang it in the place provided at the welding bench.
   c. lay it on the surface of a metal welding stool.

Instructor's Initials _____ Date _____

Things to Do

1. Practice adjusting the torch until you can maintain a neutral flame.

2. Complete the following checklist before beginning to weld:
   a. List the tools you will need for this job.

   _____

   _____

   _____

   b. List the equipment you should have ready. _____

   _____

   _____

   _____

   c. List the protective clothing you should be wearing.

   _____

   _____

   _____

   d. Check the welding area. Place a check mark (✓) in front of each item that is satisfactory.
   _____ Free of flammable materials.
   _____ Properly ventilated.
   _____ Clean and ready for use.
   _____ Clear of paints or solvents.

   e. Check the welding equipment. Place a check mark (✓) in front of each item that is satisfactory.
   _____ Oxygen and acetylene cylinder valves.
   _____ Oxygen and acetylene regulators.
   _____ Hoses.
   _____ Torch and torch valves.
   _____ Correct tip size.
   _____ Clean torch tip.
   _____ Oxygen and acetylene cylinder pressures.

3. Prepare your workpiece. Make sure it is clean, flat, and free of sharp edges and burrs. If not, describe how to properly prepare the workpiece. _____

   _____

   _____

   _____

4. When you have completed the practice piece, list any problems and their causes on the chart provided.

Problems	Causes
Example:	
Hole burned through sheet.	*Flame too close to metal.*

Unit 14

RUNNING A BEAD WITH WELDING ROD

Purpose of Welding Rod

Welding rod (also called filler metal) is used to fill a groove weld or create a fillet weld by melting or fusing additional metal into the weld joint. Welding rod is used in most oxyfuel gas welding work. *Fusion* occurs when the metal to be joined is heated and becomes molten (liquid). The molten areas of the weld joint then flow together. Welding rod may be melted and fused into the joint to provide added strength.

This unit will describe the technique for handling the rod and torch in running straight beads, a process similar to creating a continuous weld pool. See **Figure 14-1.** The difference between the two processes is that the addition of welding rod produces a convex weld bead rather than a flat or concave weld pool.

Coordinating and manipulating the welding rod and welding flame requires much practice. With experience you will be able to consistently produce beads with uniform width and height that meet American Welding Society standards.

Preparing the Equipment and Material

Before beginning the process of running a bead with welding rod, the following preparations must be made:

1. Study the practice plan, **Figure 14-2.**

2. Cut several pieces of the required material, and remove any burrs and sharp edges.

3. Clean the surface of the sheet metal with a stainless steel brush.

4. Use soapstone or chalk to draw a straight line down the center of each steel strip, **Figure 14–3.** More than one line can be drawn on each practice piece, depending on the width of the piece.

5. Position the work on a fire brick table so you can comfortably weld from right to left using the forehand method.

6. Obtain several lengths of 3/32″ (2.4 mm) diameter mild steel welding rod.

7. Select a tip size based on the thickness of the material you are welding.

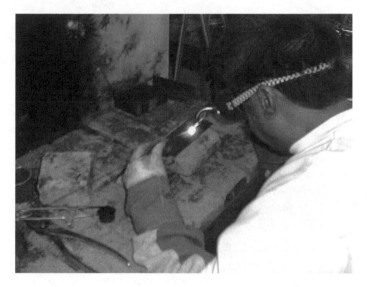

Figure 14-1. Mastering the technique of coordinating the welding rod and welding torch to run straight, evenly rippled beads takes much practice.

Procedures for Running a Bead with Welding Rod

Follow these procedures for running a bead with welding rod:

1. Set the oxygen and acetylene regulators for 6 psig (41 kPa) of pressure. Light the torch and adjust the flame to neutral.

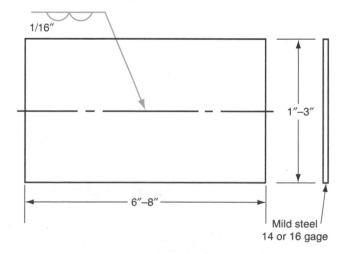

Figure 14-2. Plan for running a bead with filler metal.

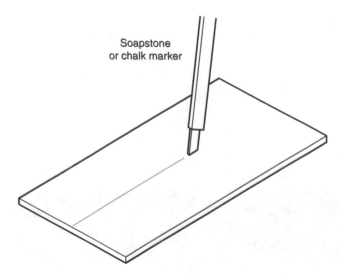

Figure 14-3. Use a soapstone or chalk marker to draw lines on material for running beads.

2. Bring the flame down to the metal surface to form a molten pool near the right-hand edge. Use a circular motion to begin the pool. Hold the torch at a 90° work angle and a 35° to 45° travel angle. (This is the same torch position used to create a continuous weld pool).

3. After the pool is formed, move the torch in a circular or semicircular pattern. Begin dipping the end of the welding rod into the molten pool. The welding rod should be held at a 90° work angle and 45° opposite the torch, **Figure 14-4.**

4. Slowly move the molten pool along the chalk line using a continuously circular motion while adding filler metal to the weld pool. Allow the flame to move around the rod to melt both the rod and base metal. A gentle up-and-down motion of the rod along the 45° axis will produce a rounded bead with an even ripple and good fusion. (Refer to Figure 14-4.) Be sure not to move the rod too far away

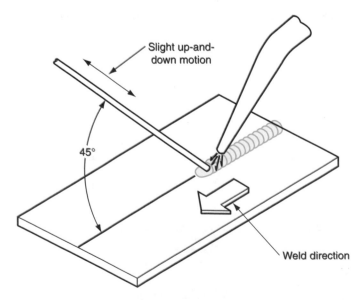

Figure 14-4. The method for adding filler metal when running a bead weld. The torch motion is the same as for creating a weld pool.

from the weld pool. It should always remain within 1/4″ of the pool so it does not cool excessively.

5. Run a continuous bead the full length of the metal. The travel speed should be slow enough to produce a uniform 3/8″ (9.6 mm) wide convex bead. If the rod becomes too short or the flame goes out, start again by forming another molten pool at the last ripple. Continue welding. **Note:** Do not throw away short pieces of welding rod. Place them end to end on the fire brick and fuse them together with the torch. Allow them to cool before using.

6. End the bead by fusing enough rod into the weld pool to keep the same bead height. Maintain a constant travel speed and rod motion to obtain uniform bead width and good fusion. The weld should penetrate approximately 25% of the metal thickness.

7. Continue running practice beads until you can consistently make them to acceptable standards. See **Figure 14-5.**

Running a Bead with Filler Metal on Thick Steel

Welding on heavy steel plate presents some new challenges. Thicker metal requires more heat to produce a quality bead. Also, running beads on thick steel requires more complex torch and rod motions. Practice is needed to properly coordinate the torch and welding rod to form acceptable beads with good penetration.

Since more heat is required, a tip with a larger orifice is needed. Select the correct tip size according to the manufacturer's specifications for the equipment used in your shop.

Procedures for Running a Bead on Thick Material

Follow these procedures for running a bead on thick steel:

1. Repeat the steps described in this unit for preparing the equipment and material. Use 1/4″ (6.4 mm) steel plate and 1/8″ or 3/16″ (3.2 or 4.8 mm) diameter mild steel welding rod.

Figure 14-5. A good example of running a bead with filler metal. Note the straightness, good penetration, and uniform bead size and shape.

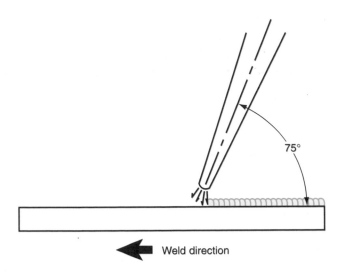

Figure 14-6. The torch is held at a 75° travel angle when welding on thick steel plate.

2. Direct the flame on the right-hand edge of the plate and form a molten pool. Add welding rod and begin welding.

3. Hold the torch at a 90° work angle and a 75° travel angle, **Figure 14-6.** Form a bead approximately 1/2″ (12.7 mm) wide. The 75° angle is used on thick steel to deliver more heat to the welding surface.

4. Because of the width of the weld pool on thick steel, weave the torch and rod as shown in **Figure 14-7.** As the torch is moved to one side of the pool, weave the rod to the opposite side.

5. Complete the weld and examine the finished bead. Check for straightness, even ripple, complete penetration, and fusion at the edge of the bead.

6. Continue running practice beads until you can consistently make them to acceptable standards. To conserve metal, use both sides of the plates for practice welding.

Check Your Progress

Write your answers in the spaces provided.

1. What is welding rod used for? _____

_____

_____

2. How is the process of running straight beads different from creating a continuous weld pool? _____

_____

_____

3. What is the technique for producing a rounded bead with an even ripple and good fusion? _____

_____

_____

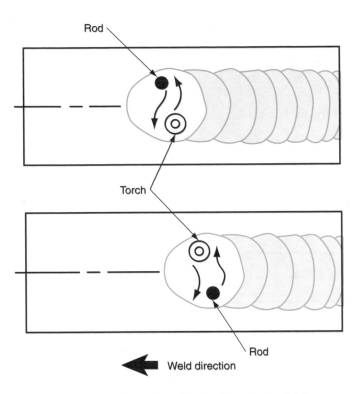

Figure 14-7. The weaving motion of the torch and rod maintains a wide weld pool and provides deep penetration.

4. What should you do if the welding rod becomes too short or the flame goes out before completing the bead weld?

_____

_____

5. Welding rod is expensive. Instead of throwing away short pieces, how can you make use of them? _____

_____

_____

6. Good penetration of the weld bead is approximately _____ percent of the metal thickness.

7. When welding thick steel plate, hold the torch at an angle of _____ with the horizontal.
   a. 15°
   b. 45°
   c. 60°
   d. 75°

8. When running a bead on thick material, move the torch to one side of the weld pool and weave the rod to the _____ side.

9. After completing the weld, examine the finished bead for the following characteristics:

   a. _____

   b. _____

   c. _____

   d. _____

10. Why should you use both sides of the plates when practicing running beads? _____

_____

_____

Instructor's Initials _____ Date _____

Things to Do

Carefully examine the bead welds you made on your practice pieces. When you have completed the practice pieces, list any problems and their causes on the chart provided. Refer to Figure 14-5 for comparison.

Problems	Causes
Example: Bead not straight.	Did not follow chalk.

Unit 15

LAP JOINT WELDING

Forming a Lap Joint

A *lap joint* is formed when two pieces of base metal are overlapped, and the bottom surface of one piece lies on the top surface of the other piece, **Figure 15-1.** The edge of one piece is fused to the surface of the second piece.

Lap joint welds present some unique challenges for the welder. The edge of the top piece melts faster and more easily than the surface of the bottom workpiece. More heat is required to melt the larger surface area of the bottom workpiece than is required to melt the smaller edge of the top workpiece.

In lap joint welding, the flame must be directed more at the surface of the bottom piece. Therefore, the torch should be positioned with a 60° work angle and a 45° travel angle, **Figure 15-2.** The tip of the flame should never be directed at the edge of the top workpiece.

To produce quality lap joint welds, the two pieces of base metal must have good contact with each other. If contact is poor, the flame can penetrate the gaps between the two workpieces, heating the underside of the top edge. This results in unnecessary melting of the edge of the top workpiece.

The two pieces of metal should be tightly clamped together. Clamping is especially important if large pieces of metal are being welded or if the lap joint is not formed in the flat position. Once the pieces are clamped, they should be tack welded at each end. A *tack weld* is a small, temporary weld used to hold the metal pieces in proper alignment to each other. For longer pieces of metal, tack welds can be made along the edge of the joint to ensure good contact and help prevent warping.

After forming the weld pool, add the welding rod halfway up the edge of the top workpiece, **Figure 15-3.** This provides additional metal on the top side and prevents overfill on the surface of the bottom piece.

Hold the welding rod just above the top edge when not dipping it into the weld pool. Welding rod should be held at a 45° angle in the opposite direction of the torch and added slowly to prevent a surplus of filler metal.

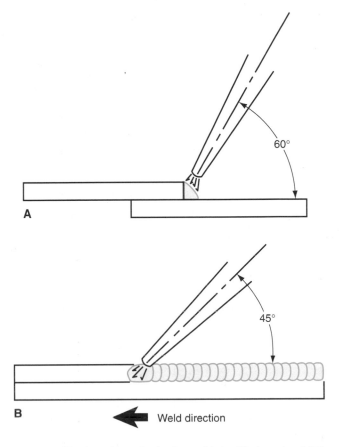

Figure 15-2. Torch positions for the lap weld. A—Work angle of 60°. B—Travel angle of 45°.

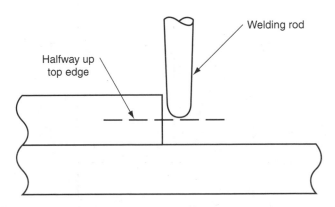

Figure 15-3. Welding rod is added halfway up the edge of the top workpiece.

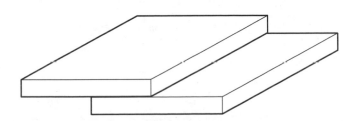

Figure 15-1. The position of thin steel workpieces for a lap joint.

Preparing the Equipment and Material

Before beginning the process of lap welding, the following preparations must be made:

1. Study the practice plan, **Figure 15-4.**

2. Cut the metal to size. Edges do not need to be specially prepared.

3. Be sure the workpieces are flat and straight.

4. Obtain 1/16″ (1.6 mm) mild steel welding rod.

5. Select the correct tip for the metal thickness.

Procedures for Welding Lap Joints

Follow these procedures for welding lap joints:

1. Position the workpieces on fire bricks. The bricks should be positioned one inch apart. The joint to be welded should be positioned over the one-inch opening. Tack weld each end of the workpieces.

2. Adjust the torch to a neutral flame. Using the proper torch angles, form a weld pool at the right-hand edge.

3. Add filler metal to the pool and begin welding using small, semicircular movements.

Figure 15-5. A completed lap weld. The weld bead has an even ripple pattern and is slightly convex.

4. Keep the weld pool in the corner of the joint as welding rod is added. Direct the tip of the flame toward the surface of the bottom workpiece.

5. Keep the bead on the top workpiece straight and uniform.

6. Inspect the weld for complete penetration, smoothness, even ripple, uniformity, and convexity. See **Figure 15-5. Note:** The face of the finished weld bead should be convex and approximately 1/4″ (6.4 mm) wide. If the weld bead is concave, too little welding rod was added or too much heat was applied. To ensure even penetration of both pieces, 1/8″ of the top workpiece should be melted away.

7. Continue practicing lap welds until they meet acceptable standards. Weld both sides of the practice pieces to save metal.

Check Your Progress

Write your answers in the spaces provided.

1. In lap joint welding, the torch should be positioned with a _____ work angle and a _____ travel angle.

2. What problem results from poor contact between the two workpieces? _____
   _____
   _____
   _____

3. The welding rod should be added (circle letter):
   a. near the outer edge of the weld pool.
   b. at the corner of the joint.
   c. halfway up the edge of the top workpiece.

4. Explain why the torch work angle should be 60° rather than 45° when lap joint welding._____
   _____
   _____
   _____
   _____

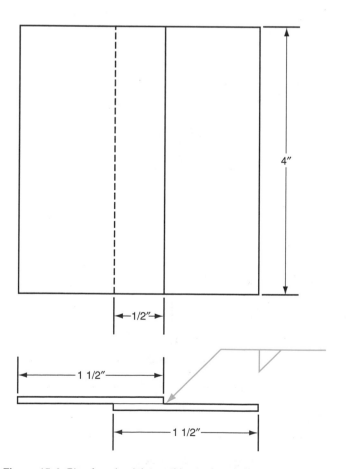

4″

1/2″

1 1/2″

1 1/2″

Figure 15-4. Plan for a lap joint on thin steel workpiece.

5. When inspecting the completed weld, look for:

    a. _____

    b. _____

    c. _____

    d. _____

    e. _____

Instructor's Initials _____ Date _____

Things to Do

1. Make a number of practice lap welds, experimenting with the torch and welding rod. Explain what happens if:

    a. Travel speed is too fast.

    _____

    _____

    b. Travel speed is too slow.

    _____

    _____

    c. Too much welding rod is added.

    _____

    _____

    d. The tip size is too large.

    _____

    _____

2. Look around your shop, at home, or in stores and make a list of items fabricated with lap welds.

3. In the space below, make a sketch of a lap joint welded on both sides.

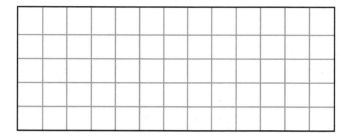

4. Explain why the basic symbol for a lap weld and fillet weld are the same. Do some research to find reference materials that will provide the answer.

5. Contact a local industry that does welding and ask for samples of lap welds. Compare these welds with the practice welds you made.

Unit 16

OUTSIDE CORNER JOINT WELDING

Forming a Corner Joint

An *outside corner joint* is formed when the edge of one piece of metal is joined to the edge of another piece at an angle of 90° or approximately 90°. Outside corner joint welds are often used in product fabrication. They can be made with or without filler material. Welding rod is not always needed when making outside corner joint welds on thin steel sheet. When the metal pieces are arranged in the correct position, excess metal from each piece serves as filler material for fusion. See **Figure 16-1.**

The torch tip is held at a 90° work angle and a 45° travel angle. The 90° work angle is critical; any other angle will melt one edge more than the other, resulting in poor uniformity and penetration.

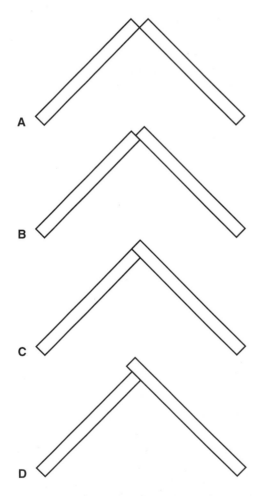

Figure 16-1. Material positions for outside corner joints. A—Open. B—Partial overlap. C—Flush. D—Overlapping.

Preparing the Equipment and Material

Before beginning the process of outside corner joint welding, the following preparations must be made:

1. Study the practice plan, **Figure 16-2.**

2. Cut the material to size, and remove any burrs and sharp edges.

3. Be sure the metal pieces are flat and straight.

4. Obtain 1/16″ or 3/32″ (1.6 mm or 2.4 mm) diameter mild steel welding rod.

5. Select the correct tip size according to the manufacturer's specifications for your equipment.

Procedures for Welding Outside Corner Joints

Follow these procedures for making outside corner joint welds:

1. Position the workpieces at a 90° angle. Make sure the inner edges touch and line up evenly.

2. Adjust the torch to a neutral flame and tack weld each end of the joint to hold the workpieces in position. See **Figure 16-3.**

3. Start a molten pool at the edge of the joint, holding the torch at a 90° work angle and a 45° travel angle.

4. Using a circular or semicircular motion, slowly move the molten pool across the joint. Do not use welding rod. Make sure the bead is formed by fusing equal amounts of metal from each workpiece into the molten pool.

5. When the weld is completed, penetration should show slightly along the inside of the joint.

6. Practice making outside corner joint welds without welding rod until you can produce consistent welds. A quality outside corner joint weld should have a uniform bead, even penetration, complete fusion, and a good surface ripple. See **Figure 16-4.**

7. Practice making several more outside corner joint welds using filler metal. A heavier, more convex bead forms when using additional filler metal.

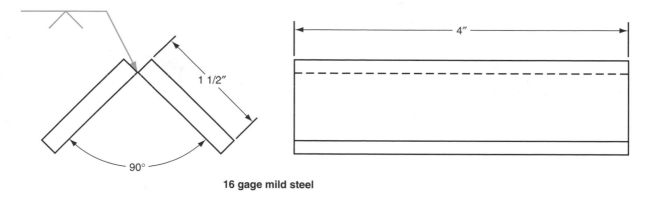

16 gage mild steel

Figure 16–2. Plan for an outside corner joint. Material can be partially overlapped.

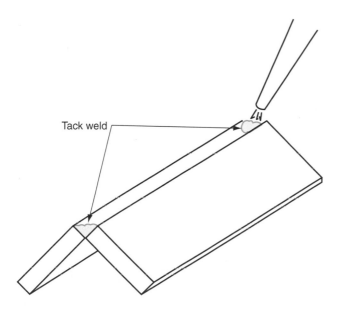

Tack weld

Figure 16–3. Material is held in position by tack welds.

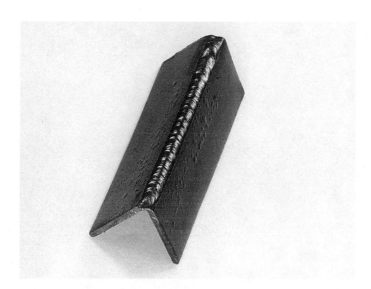

Figure 16-4. A completed outside corner joint weld on thin steel.

8. An outside corner joint weld test provides an opportunity to learn the importance of complete fusion through the cross section of the weld. Test your welds by placing each piece on an anvil with the bead up. Hammer the bead until the workpieces lie flat on the anvil surface. Examine both the face and the root of the weld bead for evidence of incomplete fusion or cracking, **Figure 16-5.** Compare the welds made with and without welding rod.

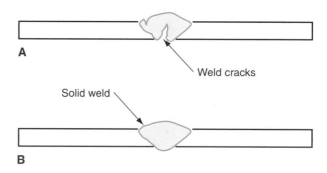

A

Weld cracks

Solid weld

B

Figure 16-5. Hammer test results. A—The weld joint is cracked at the root and face. B—A good joint with no weld failure.

Check Your Progress

Write your answers in the spaces provided.

1. Outside corner joint welds are often used in _____.

2. List the characteristics of a quality outside corner joint:

   a. _____

   b. _____

   c. _____

   d. _____

3. What is the purpose of tack welding each end of the joint to be welded?_____

   _____

   _____

4. What learning opportunity does an outside corner joint weld test provide? _____

_____

_____

5. When testing an outside corner joint weld by hammering it flat, what two things should you look for?

a. _____

b. _____

Instructor's Initials _____ Date _____

Things to Do

1. Make a list of various uses for outside corner joints in industry. Gather the information from reference materials in your shop or library.

2. Set up and weld an outside corner joint with the workpieces in the *flush position*. Refer to Figure 16-1. In the space provided, list any problems you encountered.

_____

_____

_____

_____

_____

3. Set up and weld an outside corner joint with the workpieces in the *overlapping position*. Refer to Figure 16-1. In the space provided, list any problems you encountered.

_____

_____

_____

_____

_____

4. Keep some of your practice pieces that were not used for the hammer test. Use them to practice fillet welding on an inside corner. Making fillet welds on inside corner joints and T-joints is introduced in Unit 17.

Unit 17

T-JOINT WELDING

Forming T-joints

A **T-joint** is formed when the edge of one piece of metal is joined to the surface of another piece of metal at a 90° angle. The T-joint is welded in much the same way as the lap joint. A **fillet weld** is used to join the two surfaces of the T-joint. The fillet weld is also used to weld lap joints and inside corner joints. When making a fillet weld on a T-joint, the heat must be equally distributed between the two steel workpieces.

The two workpieces may be arranged in either a horizontal or flat position, **Figure 17-1.** The flat position makes welding much easier because the torch can be held straight up and down at a 90° work angle.

The flat position is most efficient for making fillet welds on a T-joint, **Figure 17-2.** Holding the torch at a 90° work angle and a 60° travel angle, while directing the flame tip at the center of the joint, ensures both workpieces are heated equally by the flame. Flat position welding forms a trough into which the weld bead can be symmetrically shaped as it is moved along the T-joint.

Making fillet welds in the horizontal position is more difficult than in the flat position. Even so, welding T-joints in the horizontal position is widely used, so you will need much practice in this technique. See **Figure 17-3.**

Undercutting the vertical workpiece is a common problem when first attempting to weld a T-joint in the horizontal position. See **Figure 17-4.** *Undercutting* refers to a depression formed in the base metal when it is melted away below the original surface level. Undercutting is caused by applying too much heat, holding an improper torch angle, or failing to add filler metal in the proper place.

Figure 17-2. Making a fillet weld on a T-joint in the flat position. Flat position welding is the easiest and most efficient position.

Figure 17-3. The welder is using a circular torch motion and adding welding rod to make a fillet weld on a T-joint in the horizontal position.

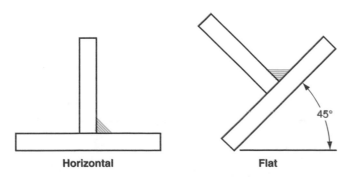

Figure 17-1. T-joints set up in the horizontal and flat positions.

Horizontal

Flat

45°

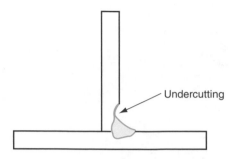

Figure 17-4. Undercutting is caused by improper torch angle, excessive heat, or failure to add filler metal to the topside of the weld pool.

Preparing the Equipment and Material

Before beginning the process of welding a T-joint, the following preparations must be made:

1. Study the practice plan, **Figure 17-5.**

2. Cut the metal pieces to the required size.

3. Be sure the edges of the vertical pieces are flat and smooth. It is important the vertical piece fits flush against the surface of the horizontal piece.

4. Obtain 3/32″ (2.4 mm) diameter mild steel welding rod.

5. Select the correct size tip for the metal thickness. A tip one size larger than specified may work better after a few practice welds.

Procedures for Making Fillet Welds

Follow these procedures for fillet welding:

1. Position one workpiece flat on the firebricks.

2. Using pliers, hold the vertical workpiece in the center of the flat workpiece. Tack weld each end. The workpieces are now set up in the horizontal position.

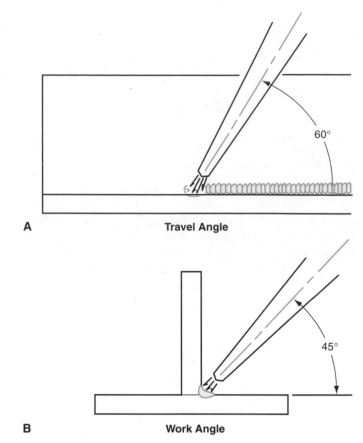

Figure 17-6. Torch angles for fillet welding. A—Travel angle of 60°. B—Work angle of 45°.

3. If the workpieces are not at right angles to each other after making the tack welds, slightly bend the vertical workpiece until it is in correct alignment.

4. Set both the oxygen and acetylene pressure regulators to 6 psig (21 kPa). Light the torch and direct the flame at a 45° work angle and a 60° travel angle, **Figure 17-6.** Carefully form a molten weld pool in the corner at the right-hand side of the joint.

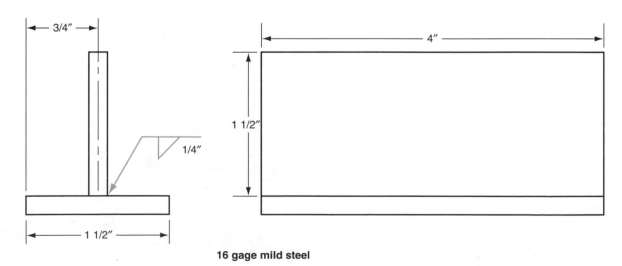

16 gage mild steel

Figure 17-5. Plan for a fillet weld on thin steel.

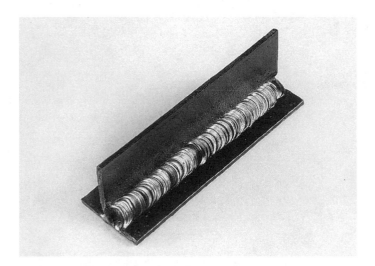

Figure 17-7. Completed fillet weld on thin steel.

5. Add welding rod and begin welding using the forehand technique. Use a circular torch motion. Keep the tip of the flame approximately 1/8″ (3.2 mm) away from the surface. Heating the vertical workpiece too much will cause undercutting. This can be overcome by keeping the weld pool moving and adding rod to the top of the molten pool. The welding rod will help draw heat away from the vertical workpiece.

6. As you weld, add enough welding rod to produce a 1/4″ to 3/8″ (6.4 mm to 9.6 mm) bead width with 1/4″ (6.4 mm) equal legs. Always keep the molten pool centered in the corner of the joint.

7. Practice making fillet joint welds until you can consistently make uniform beads. Use both sides of the vertical workpiece, but allow the first bead to cool before welding the opposite side.

8. Visually inspect the finished fillet weld, **Figure 17-7.** Check the legs for equal length. Examine the width and ripple pattern of the weld bead for uniformity along the entire length of the weld.

Check Your Progress

Write your answers in the spaces provided.

1. A fillet weld is used to join the two surfaces of a T-joint. The fillet weld is also used to weld _____ joints and _____ joints.

2. List two reasons it is easier to make quality fillet welds in the flat position than in any other position:

   a. _____

   b. _____

3. A problem that typically arises when first attempting the fillet weld is _____ the vertical workpiece.

4. When welding a T-joint, how far should the inner cone of the flame be kept from the surface? _____

5. The finished bead width on a 16-gage T-joint should be _____ with _____ equal legs.

Instructor's Initials _____ Date _____

Things to Do

1. Make fillet welds on both sides of a T-joint without allowing the first side to cool. Explain the difference in the amount of heat required for the second weld compared with the first.

2. In the space below, sketch a cross section of a fillet weld made on both sides of a T-joint. Draw the correct welding symbol for this type of welded joint.

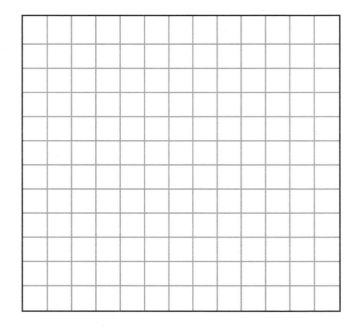

3. Visit a local welding manufacturer and obtain samples of properly made fillet welds.

Unit 18

BUTT JOINT WELDING

Forming a Butt Joint

A **butt joint** is used to join two pieces of metal end to end or side to side. The butt joint weld is one of the most commonly used joints in oxyfuel gas welding. On thin steel, the butt joint needs to be welded from one side only since full penetration to the root of the weld can be achieved in a single weld pass. Thin steel is less than 3/16″ (4.8 mm) in thickness. Special joint preparation is not needed in butt welding thin steel. The edges need only be squared off and straight.

On thick steel, butt joint welding requires special joint preparation. Several weld passes must be made to achieve complete fusion between the workpieces and to fill the groove. (Unit 25 covers butt welding thick steel.)

Mastering the technique of running straight, uniform beads with welding rod is essential to making quality butt joint welds. Controlling distortion as the metal is heated is a concern in butt joint welding. The gap between the two workpieces tends to narrow as the weld continues. Placing tack welds at each end of the workpieces before beginning the weld will maintain alignment, **Figure 18-1.**

Preparing the Equipment and Material

Before beginning the process of butt welding, the following preparations must be made:

1. Study the practice plan, **Figure 18-2.**

Figure 18-1. A U-shaped spacer rod is inserted in the root opening. Tack welds are made at each end of the joint to maintain alignment.

2. Cut or shear the metal to size. Use a file or a grinder to remove sharp edges and square the edge.

3. Be sure the metal is flat.

4. Obtain a 1/16″ or 3/32″ (1.6 or 2.4 mm) diameter mild steel welding rod.

5. Select the correct tip size for the metal thickness.

Procedures for Welding Butt Joints

Follow these procedures for butt welding:

1. Place the workpieces flat on the firebricks. Position the workpieces side by side with a 1/8″ (3.2 mm) root opening made with a U-shaped 1/8″ (3.2 mm) spacer wire. (Refer to Figure 18-1.)

2. Tack weld the ends of the workpieces together. To make the tack welds, apply a neutral flame to one end of the joint until the corners of both workpieces turn cherry red. When the corners begin to melt, add the welding rod to bridge the 1/8″ (3.2 mm) gap. Continue adding welding rod until a 1/4″ to 1/2″ (6.4 mm to 12.7 mm) long tack weld has been deposited. After making the first tack weld, use the U-shaped spacer wire to realign the root opening. Make sure the gap is still 1/8″ (3.2 mm). Then, make the second tack weld at the other end of the joint.

3. After tack welding is complete, align two firebricks side by side with a 1″ (25 mm) gap between them. Place the workpiece on the firebricks with the 1/8″ (3.2 mm) root opening centered over the 1″ (25 mm) gap.

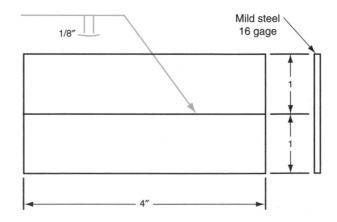

Figure 18-2. Plan for a butt joint on thin steel.

Figure 18-3. A keyhole shape is formed by evenly melting the edge of each workpiece.

4. Begin preheating the workpieces from the right side. A side-to-side or circular torch motion helps to melt the edges of both workpieces. Use the same torch and rod motions you used when running a bead. The torch should be held at a 90° work angle and a 35° to 40° travel angle.

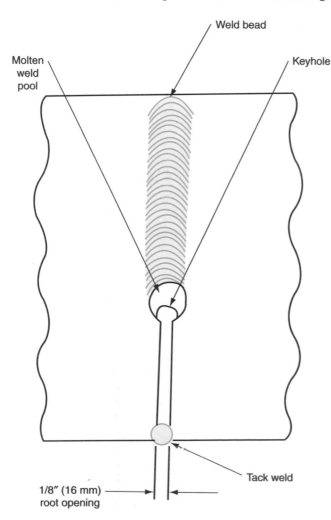

Figure 18-4. Top view of a properly formed keyhole.

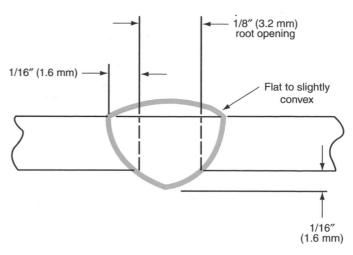

Figure 18-5. Cross section of a finished butt joint weld on thin steel workpiece.

5. Hold the torch and welding rod centered over the square groove. This will keep the weld bead straight. Evenly melt the edges of each workpiece through to the bottom side of the workpieces. When done correctly, a wider molten hole (approximately 3/16″ or 4.8 mm) will form the shape of an old-fashioned keyhole, **Figure 18-3.** This keyhole shape should be maintained at all times to ensure complete penetration of the weld, **Figure 18-4.**

6. When the edges of the pool are molten, add the welding rod at the leading edge of the keyhole. Move the torch slowly along the joint to create a weld slightly over 1/4″ (6.4 mm) wide. The finished weld face is flat to slightly convex, while the root penetration forms a 1/16″ (1.6 mm) convex bead on the bottom side of the workpiece, **Figure 18-5.**

7. Practice butt welding until your joints meet acceptable standards, **Figure 18-6.**

8. Visually inspect the weld for bead straightness, uniformity of ripple and width, and complete penetration on the back side. If the weld face is concave rather than flat, too much heat was applied, or the travel speed was too slow, allowing the weld to sag.

Figure 18-6. Completed butt weld on thin steel workpiece.

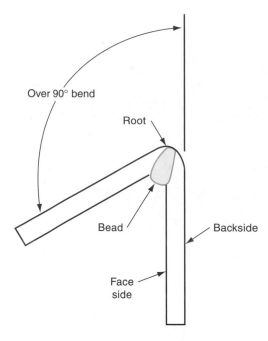

Figure 18-7. Destructive bend test for a butt weld on thin steel.

9. Test the butt weld using a destructive evaluation test. Place the welded piece in a vise with the bead just above the vise jaws. Hammer the backside of the weld workpiece until the bend surpasses a 90° angle, **Figure 18-7.** An acceptable weld will not show any cracks along the root of the bead.

Check Your Progress

Write your answers in the spaces provided.

1. On thin steel, a butt weld needs to be welded on _____ only.

2. A major concern in butt welding is *(circle letter)*:
   a. beginning the weld.
   b. keeping the bead straight.
   c. maintaining an even ripple.
   d. controlling distortion of the metal.

3. Tacking the pieces before beginning to weld will help maintain _____.

4. For butt welding thin steel, use the same torch and rod motion used when _____.

5. List three things to look for when visually inspecting a butt weld.

   a. _____

   b. _____

   c. _____

Instructor's Initials _____ Date _____

Things To Do

1. Compare your practice butt welds with samples supplied by your instructor. Make a list of any similarities and differences you find in your welds.

2. Take a field trip to a local welding shop. Ask an experienced welder to give your class a demonstration on butt welding thin and thick steel workpieces.

3. Check the standards of acceptability for weld soundness prepared by the American Welding Society. Compare your practice welds with these standards. In the space below, list any variations between your welds and those acceptable by AWS. _____

   _____

   _____

   _____

   _____

   _____

   _____

4. Write to a commercial welding company and ask for a list of the types of tests a welder must pass to be hired.

Unit 19

OUT-OF-POSITION WELDING

Basic Positions

Four basic positions are used in welding: flat, horizontal, vertical, and overhead, **Figure 19-1.** The figures in this book mainly depict flat position welding. It is easy, efficient, and the position of choice when the parts are able to be moved and aligned. However, many welding jobs require a different position because the parts are fixed and cannot be moved into the flat position. *Out-of-position welding* refers to welding done in a position other than flat.

Out-of-position welding is more difficult than flat position welding, so the welder must have more training and practice to weld out-of-position. Welds made in the horizontal, vertical, and overhead positions must meet the same high standards as welds made in the flat position. In flat position welding, the welder's hands and elbows can rest on the working surface. However, out-of-position welding is often done at or above eye level with the arms held above the shoulders. Holding the torch and welding rod steady requires substantial physical strength. Welders can easily become fatigued and mistakes can readily occur. However, welders who have proven their skill in out-of-position welding can increase their income.

This unit will focus on welding in the horizontal, vertical, and overhead positions. Examples of welding various joints in certain positions will be provided. Butt joints in the horizontal and vertical positions, and the T-joint in the overhead position will be covered. These are some of the more difficult welds to make. Any weld joint can be made in any position.

Warning: Flame-resistant leather clothing should be worn in out-of-position welding. Button the collars of your shirt and welding jacket to prevent hot metal or slag from falling onto your skin. Always wear a welder's cap when doing overhead welding. Welding goggles or a welding helmet with #4 to #6 lenses are required.

Preparing the Equipment and Material

Before lighting the torch and practicing out-of-position welding, make all the necessary preparations for welding

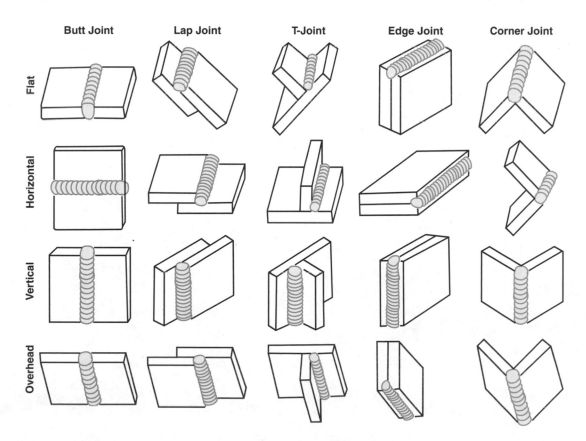

Figure 19-1. The five most common weld joints may be welded in any of the four welding positions.

outlined in previous units. Select the correct torch tip and welding rod based on metal thickness and the manufacturer's recommendations.

Decide what type of joint you want to practice; then, prepare the material according to the plan provided in one of the previous chapters. For example, if you want to practice out-of-position butt welding, review the plans in Unit 18 for setting up a butt joint. After cutting the metal workpieces to the correct size, prepare the edges and tack weld the pieces together in the flat position. Next, clamp the practice piece in a welding fixture in the position you want to practice.

Welding in the Horizontal Position

In out-of-position welding, gravity is always working to deform the weld bead. To compensate for this downward pull, two techniques are especially helpful. First, adjust the torch so it points slightly upward. This enables the force of the welding gases to push the molten metal upward and help overcome the pull of gravity. Second, place the welding rod into the upper edge of the weld pool. This helps to create an evenly centered weld bead by compensating for the sagging of the molten metal. Hold the welding rod at a 30° to 40° angle opposite the torch and 10° up from the horizontal.

Figure 19-2 shows a welder making a butt joint in the horizontal position. The travel angle should be 70° to 80°, as shown in the top view of **Figure 19-3**. The torch tip and flame should be directed upward forming a 10° to 30° work angle, as shown in the front view of Figure 19-3. These torch angles can be adjusted as long as the edges of both metal workpieces are evenly heated and melted to form an acceptable weld bead.

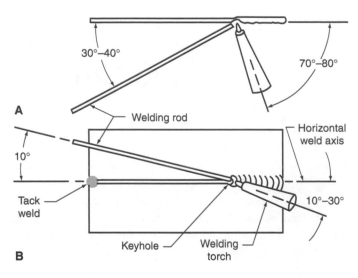

Figure 19-3. Welding torch and rod angles for a butt joint in the horizontal position. A—Top view. B—Front view.

Welding in the Vertical Position

Most welders would agree the vertical welding position is not as difficult as the horizontal and overhead positions. Gravity is working equally and evenly on both edges of the joint being welded. In the horizontal position, however, gravity is pulling the molten metal from the top edge toward the bottom edge.

Vertical welds are usually started at the lowest point on the joint and moved upward by the forehand method. Working from bottom to top, the welder can easily see the weld pool and control it by pushing it upward. The upward pressure of the welding gases helps to keep the weld pool from sagging. Vertical welds can be made from the top down, but the weld pool is more difficult to control since it tends to run down the joint.

Figure 19-4 shows a welder making a butt joint in the vertical position. The torch should be positioned with a 90° work angle and a 30° to 40° upward travel angle. Hold the

Figure 19-2. A butt joint in the horizontal position. The torch is angled slightly upward to compensate for gravity. Note the welder is left-handed.

Figure 19-4. A butt joint in the vertical position is welded from bottom to top. The force of the gases helps to push the molten weld pool upward.

Figure 19-5. Welding a T-joint in the overhead position. Overhead welding is done at or above eye level and requires patience and physical strength.

welding rod opposite the torch at a 30° angle from the surface of the workpiece. On thin metal, the torch can be kept centered in the middle of the joint. Hold the torch steady as you travel upward. On thick metal, a side-to-side weaving motion is required. Be sure to equally heat the edges of both metal workpieces so the weld pool is centered. To ensure complete penetration to the root of the joint, dip the welding rod halfway into the leading edge of the weld pool.

Welding in the Overhead Position

Overhead position welding is similar to flat position and horizontal position welding in that the same torch and welding rod angles are used. The difference is the angles are rotated 180° into an overhead position. The challenging part of overhead welding is holding the torch and welding rod steady while tilting your head up to view the weld. The forehand welding technique is used in overhead welding, meaning right-handed welders work from right to left.

The key to working in the overhead welding position is getting in a comfortable body position that provides a good view of the weld. Welders might choose to stand in line with the weld axis or across from the weld axis. Standing in line with the weld axis gives the best view of the weld pool, which is especially helpful when welding a butt joint. Standing across from the weld axis works well when welding a T-joint or a lap joint, **Figure 19-5.**

When welding a T-joint in the overhead position, the torch should be positioned with a 45° work angle and a 30° to 45° travel angle. See **Figure 19-6.** Hold the welding rod at a 15° to

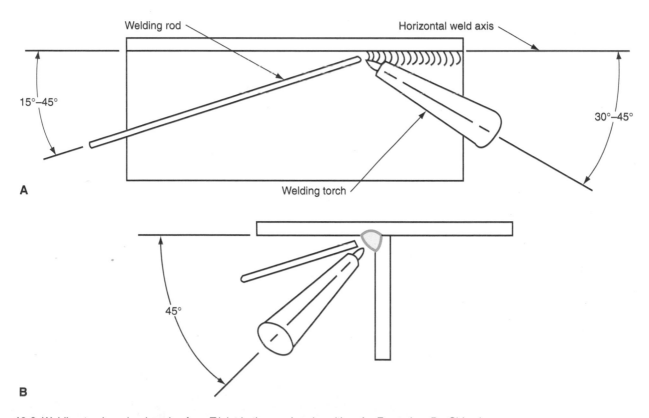

Figure 19-6. Welding torch and rod angles for a T-joint in the overhead position. A—Front view. B—Side view.

45° angle opposite the torch. Keep the torch steady as you smoothly move it along the joint.

Making a fillet weld in the overhead position is similar to welding in other positions. Wait for a C-shaped molten weld pool to form before adding the welding rod. When welding overhead, add the welding rod near the upper part of the weld pool. Gravity will pull the molten metal down onto the vertical piece.

Remember: Undercutting can result from too much heat, holding the torch at the wrong angle, or adding the welding rod in the wrong place. Undercutting weakens the base metal.

Check Your Progress

Write your answers in the spaces provided.

1. What are the four basic welding positions?

    a. _____

    b. _____

    c. _____

    d. _____

2. What does *out-of-position welding* mean? _____

    _____

3. Name two safety tips to keep in mind when doing overhead welding.

    a. _____

    b. _____

4. What two techniques are used to compensate for gravitational pull on the molten weld pool when making a horizontal weld?

    a. _____

    _____

    b. _____

    _____

5. The torch and welding rod angles used in the overhead welding position are the same as those used in the _____ and _____ positions, except the angles are rotated 180°.

Instructor's Initials _____ Date _____

Things to Do

1. Follow the plan in Unit 15 for setting up the lap joint. Practice welding lap joints in the horizontal, vertical, and overhead positions. Continue practicing lap joint welds in each position until they meet acceptable standards.

2. Follow the plan in Unit 17 for setting up the T-joint. Practice welding T-joints in the horizontal, vertical, and overhead positions. Continue practicing fillet welds on T-joints in each position until they meet acceptable standards.

3. Follow the plan in Unit 18 for setting up the butt joint. Practice welding butt joints in the horizontal, vertical, and overhead positions. Continue practicing butt joint welds in each position until they meet acceptable standards.

Unit 20

INSPECTION OF WELDS

Flaws and Defects

Every completed weld should be inspected by the welder. Inspection enables the welder to locate flaws or defects in the weld. It also provides important information about corrections to be made to future welds. Finally, inspection helps the welder or the inspector determine where a weld may need repair.

Every weld has flaws that may or may not render the weld unusable. A *flaw* is an imperfection in the weld, such as a place where the weld was stopped then restarted. Many flaws can be visually detected by a skilled welder, **Figure 20-1.** Visible imperfections include:

- Small cracks.
- Lack of penetration.
- Undercut (a depression where the weld has cut below the surface of the base metal).
- *Porosity* (tiny bubbles where gas was trapped in the base metal).
- *Inclusions* (unwanted foreign material in the weld).
- Spattering.
- Excessive *slag* (metal oxide formed on the underside of flame cut steel).

Other welding flaws are incorrect size, width, contour, and location of the weld. **Figure 20-2** shows fillet welds that would and would not pass visual inspection.

Locating flaws is just one purpose of inspection. Identifying defects in the weld is another. A *defect* is a flaw that renders the weld unusable unless it can be repaired. A weld with a defect fails to meet the proper specifications for that weld job.

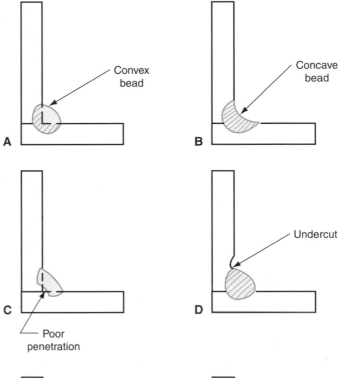

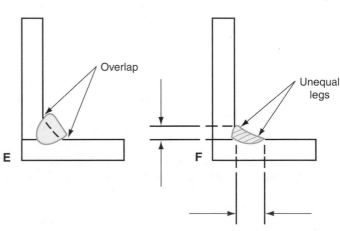

Figure 20-2. Visual inspection of fillet welds. A and B—The convex and concave beads show good penetration and would pass visual inspection. Views C through F show common flaws that can be detected visually. C—Poor penetration. D—Undercut. E—Overlap. F—Unequal legs.
Note: Poor penetration is only visible on the ends of a joint. X-ray and ultrasonic inspection can locate poor penetration in a fillet weld.

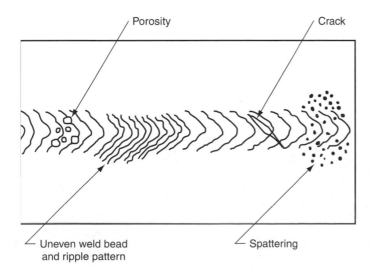

Figure 20-1. Various flaws and defects can be easily detected by doing a visual inspection.

The allowable size of a defect depends on the job being done. Depending on the specifications, a small crack could be a flaw on one weld job and a defect on another. For example, if a small crack developed while making a weld on a high-pressure natural gas pipeline, the crack would be considered a defect. The weld would be unusable and would not pass inspection. If the same size crack developed while welding a steel frame for a small worktable, the crack might be considered a flaw and could pass an inspection.

Many tests can be done to locate flaws and defects. These tests fall into two categories of evaluation: nondestructive and destructive. *Nondestructive evaluation* does not damage the weld or the metal piece in any way. *Destructive evaluation* makes the weld unfit for use after the test is finished.

Nondestructive Evaluation

Many types of inspections can be done on a weld without damaging it. Experienced welders can easily spot most problems with a *visual inspection.* This is the most common type of nondestructive test. Other types of nondestructive tests require expensive equipment. Other types include x-ray, ultrasonic, liquid penetrant, and magnetic particle inspections and air- or water-pressure leak tests.

Beginning welders only need to perform visual inspections to make sure the weld is the correct size. Check the weld against a drawing, print, or written plan that indicates the proper dimensions for the job. Tape measures, calipers, and other measuring tools can assist the welder and inspector in determining if the weld was made to the correct dimensions.

A trained eye can quickly locate flaws or defects on the surface of the weld. Visual inspection is a fast, easy, and accurate way of determining if a weld is acceptable.

A visual inspection is also helpful in evaluating the welder's skills and techniques. A visual inspection can show if travel speed was too slow or too fast, torch angles were correct, the flame was properly adjusted to provide the correct amount of heat, welding rod was added at the proper rate, and other clues. The inspection helps determine where improvement is needed.

Destructive Evaluation

Destructive evaluations cause the weld to be damaged or completely destroyed. Such inspections are used to determine the physical properties of a weld, such as its strength.

Destructive evaluation of welds usually requires testing equipment. Some common destructive tests done in large industry include the tensile test (to determine tensile strength and ductility), bend test, fillet test, hardness test, peel test, and pressure test. The fillet and bend tests can be done in the shop without expensive equipment.

Fillet Test

The *fillet test* is a destructive test used to determine the quality of a fillet weld. A workpiece with a fillet weld is placed in position on an anvil, **Figure 20-3.** Force is applied with a hammer until the vertical piece is bent flat against the

horizontal piece or until the joint breaks. If the weld breaks, it should break along the center axis of the weld bead. This means penetration is good and the legs are equal.

If the weld breaks along one edge or the other, poor fusion or uneven fusion along the edges is indicated. If the legs of the weld are uneven, it may mean either the torch angle or torch position was incorrect. Check the fractured surface for porosity, inclusions, or poor penetration.

If the broken or bent surface shows no signs of porosity, inclusions, or other defects, the weld will pass inspection. If the weld does not break when the vertical piece is bent flat against the horizontal piece, a strong weld is indicated and it should pass inspection.

Bend Test

The *bend test* is a destructive test for accurately determining the quality and ductility of a butt weld. A sample test piece is bent into the shape of a "U," using either a specialized hydraulic machine or the force of a hammer.

Before doing the bend test, a sample test piece must be prepared. Typically, the butt weld to be tested is cut into 1″ to 1 1/2″ (25.4 to 38.1 mm) wide strips across the weld. After the strips are cut, grind down the weld bead on the face and root side so it is flush with the rest of the plate. **Note:** Prior to grinding and bending, mark each side of the strips with an "F" or "R" to indicate the face side and root side of the weld.

Place the strip in a vise with the face of the weld toward you. Bend the strip away from you by hammering it to at least a 90° angle. If the weld does not break, continue to bend it by squeezing the plate in a vise until a parallel U-shape is formed. Inspect the weld for defects. This is called an *unguided bend test.*

A *guided bend test* forces the strip through a set of dies to create a specific radius on the sample. See **Figure 20-4.** The guided bend test is more repeatable than the unguided bend

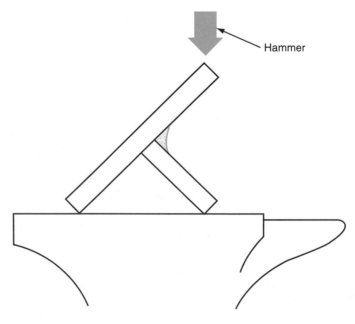

Hammer

Figure 20-3. Position of assembly for hammer-testing the finished fillet weld.

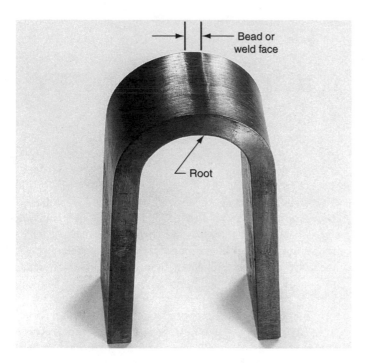

Figure 20-4. Butt weld test piece shows no cracking or porosity along the face or edge of the weld.

test. If the weld breaks at any point in the bending process, it fails the test. A quality weld will bend to a U-shape without showing any signs of cracking or porosity. Flaws at the edge of the sample are usually not counted. To inspect the root side, repeat the procedure, but place the next strip in the vise with the root of the weld toward you.

Check Your Progress

Write your answers in the spaces provided.

1. List seven flaws that can be quickly detected with a visual inspection.

   a. _____

   b. _____

   c. _____

   d. _____

   e. _____

   f. _____

   g. _____

2. Explain the difference between a flaw and a defect.

   _____

   _____

   _____

3. Explain the difference between destructive and nondestructive evaluations. _____

   _____

   _____

   _____

4. The _____ test is used for inspecting butt joint welds, while the _____ test is used for inspecting fillet joint welds.

5. Fill in each blank with the name of the common flaw described:

   a. _____ is a depression where the weld has cut below the surface of the base metal.

   b. _____ is a weakness caused by tiny bubbles where gas was trapped in the base metal.

   c. _____ are unwanted foreign material in the weld.

Instructor's Initials _____ Date _____

Things to Do

1. Make a practice fillet weld on one side of a T-joint. Test the weld by hammering it until it breaks or is completely flattened. Examine the workpiece and list any defects. Explain how the defects may have developed.

2. Make a practice butt weld following the procedure in Unit 18. Prepare a sample test piece and conduct an unguided bend test. Examine the weld and list any defects.

3. Write to a commercial welding company, and ask for a list of the types of tests a welder must pass to be hired.

4. Ask your instructor to evaluate your welds for flaws. List the reasons these flaws occurred and the techniques that need improvement to eliminate each flaw.

5. Take a field trip to an industrial plant that conducts nondestructive and destructive tests on weld pieces to view the evaluation process firsthand.

Unit 21

CUTTING EQUIPMENT

Cutting Equipment versus Welding Equipment

Oxyfuel gas cutting equipment is used throughout the metal-working industries for cutting steel and ferrous (iron-based) metals. The equipment is basically the same for both oxyfuel gas cutting and oxyfuel gas welding, except for the torch, **Figure 21-1.** The cutting torch is a unique and invaluable tool that enables the operator to cut almost any thickness of steel to the required size and shape.

Besides the torch, the oxygen regulator may differ between the oxyfuel gas welding outfit and cutting outfit. A high flow of oxygen is necessary to cut very thick metal. Therefore, a heavy-duty oxygen regulator with high-pressure indications and large-volume capacity may be needed.

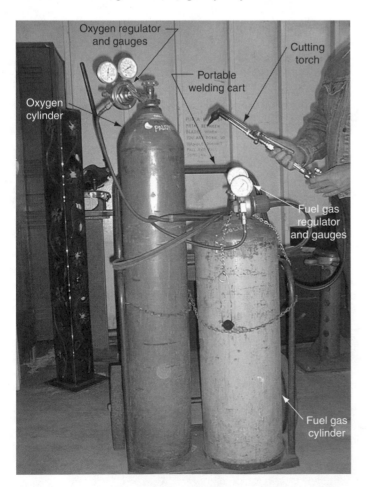

Figure 21-1. The oxyfuel gas cutting outfit is the same as the welding outfit, except for the torch and a heavy-duty oxygen regulator.

Oxyfuel gas cutting is also referred to as flame cutting. *Flame cutting* is the process of cutting steel by removing material to form a slot, called the *kerf.* The metal is actually removed from the kerf by burning it away. Flame cutting is one of the fastest and most efficient methods for cutting thick metal.

Essential Elements for Burning

To understand how metal can be cut quickly with a flame, you must first understand how burning takes place. Three essential elements are needed for burning: fuel, oxygen, and heat. Fuel can be any material that will chemically combine with oxygen. In this case, the steel to be cut *is* the fuel.

Second, a large quantity of pure oxygen is needed to combine with the fuel. The more oxygen present, and the greater the oxygen's force, the faster burning will take place. Therefore, flame cutting requires a jet stream of pure oxygen directed onto the steel.

Third, heat is needed to raise the temperature of the steel until it reaches its ignition temperature. *Ignition temperature* is the temperature at which the material will burn when oxygen is present. The ignition (kindling) temperature of steel is 1500°F (816°C). At this temperature, steel turns cherry red. The stream of oxygen is directed onto the hot metal. The steel burns rapidly, and a slot is cut through the entire piece. The oxygen and hot steel are actually combining to form a chemical reaction called *oxidation.*

A variety of gases can be used for oxyfuel gas cutting. Acetylene and MAPP (methylacetylene-propadiene) gases are the most widely used. Propane, natural gas, and hydrogen are also used.

The Cutting Torch

The oxyfuel gas cutting torch is an ideal tool for accurately controlling the heating and cutting action, **Figure 21-2.** The torch is designed to preheat the steel until it reaches the proper temperature. Then, the torch supplies a stream of oxygen to begin the burning (or oxidizing) process. The stream of oxygen that cuts the metal is referred to as *cutting oxygen.* The oxygen is released by pressing the *cutting oxygen lever.*

A cutting torch is similar to a welding torch. It provides a method of mixing oxygen and acetylene in proper proportions to produce the desired preheating flame. However, the cutting torch has two additional features:

- A third passageway for high-pressure cutting oxygen, which is regulated by the cutting oxygen lever on the torch, **Figure 21-3.**
- A special cutting tip.

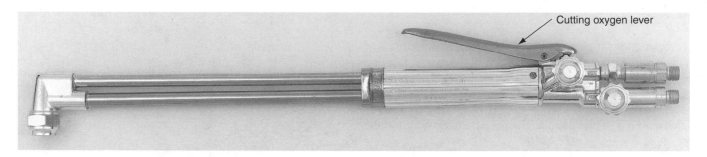

Figure 21-2. A standard oxyfuel gas cutting torch. (Uniweld Products, Inc.)

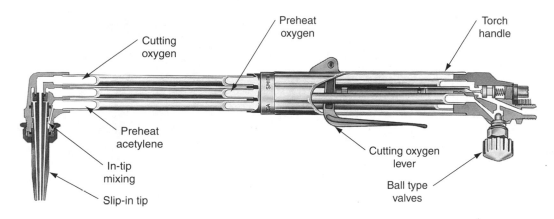

Figure 21-3. Construction of the cutting torch. (Smith Welding Equipment Co.)

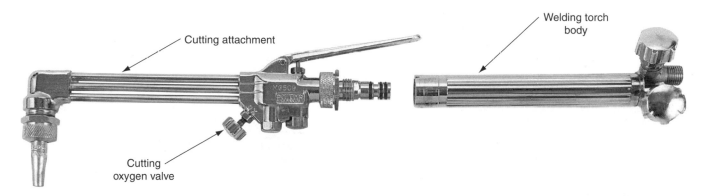

Figure 21-4. A cutting attachment may be connected directly to the welding torch body. (Smith Welding Equipment Co.)

Two types of cutting torches are widely used. The first is a standard, one-piece cutting torch used only for cutting. The second is called a cutting torch attachment or two-piece combination torch.

The welding torch can be converted to a cutting torch by removing the welding tip from the torch body and directly fastening the cutting torch attachment, **Figure 21-4.** Notice the cutting torch attachment has an extra cutting oxygen valve. (The procedure for opening and closing the extra valve is described in Unit 22.) The advantage of the cutting torch attachment is that it can be quickly added and removed as needed, **Figure 21-5.** The cutting torch, on the other hand, must be connected to flashback arrestors and hoses before use.

Figure 21-5. A standard cutting torch attachment that quickly attaches to a torch body. (Uniweld Products, Inc.)

The Cutting Tip

Cutting torch tips are made of copper and are longer than welding tips, **Figure 21-6.** The diameter of the end of a cutting tip is much larger than the end of a welding tip. The cutting tip has a large orifice in the center through which the high-pressure oxygen flows. This opening is surrounded by smaller orifices, often in a circular pattern of four to eight holes for the preheating flames, **Figure 21-7.**

The size of the cutting tip is determined by the thickness and type of carbon or alloy steel to be cut. Cutting tips are designed and specified to provide the correct volume of heating and cutting gases for particular metal thicknesses. A wide variety of customized cutting tips is available for specific cutting jobs, **Figure 21-8.**

Each cutting tip manufacturer has a different numbering system for identifying the type and style of tip. Always check the manufacturer's specifications to select the correct cutting tip. The manufacturer provides a chart with specific data on metal thickness, tip size, regulator pressure settings, and travel speed.

Safety Precautions

Flame cutting can be very hazardous if proper precautions are not taken. Always wear goggles with suitable lenses to protect your eyes from the intense light and flying sparks. Wear a welding cap to cover your hair. Gauntlet welding gloves and a long leather apron or welding outfit are needed to protect the body from hot slag.

Explosive or flammable materials must be removed from the cutting area. Red hot particles of molten slag can splatter great distances from the high-pressure stream of oxygen. A fire extinguisher should be readily available at all times.

Cutting operations should be performed in an area away from other persons to protect them from flying slag. Check equipment, protective clothing, and the cutting area before lighting the torch.

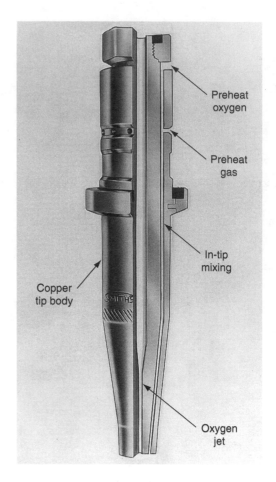

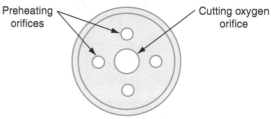

Figure 21-7. Cutting torch tip. A—Details of a typical cutting tip. (Smith Welding Equipment Co.) B—Location of preheating and cutting oxygen holes.

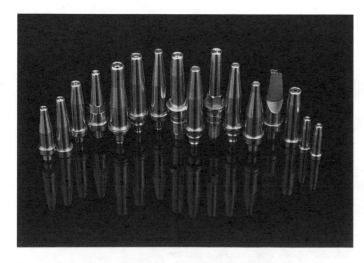

Figure 21-6. A wide variety of cutting tip sizes is available. (American Torch Tip Co.)

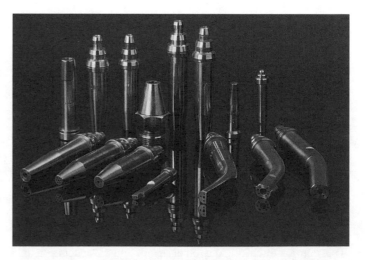

Figure 21-8. Assortment of specialty cutting tips. (American Torch Tip Co.)

Check Your Progress

Write your answers in the spaces provided.

1. What three elements are needed for burning to take place?

   a. _____

   b. _____

   c. _____

2. Steel can be cut when it reaches an ignition temperature of *(circle letter)*:
   a. 1100°F (593°C).
   b. 1500°F (816°C).
   c. 2200°F (1205°C).

3. Name the two types of cutting torches, and explain how they differ.

   a. _____

   _____

   b. _____

   _____

4. A cutting tip has a center hole for high-pressure oxygen surrounded by smaller orifices for the _____.

5. Selecting a cutting tip of the correct size is primarily based on:

   a. _____

   b. _____

Instructor's Initials _____ Date _____

Things to Do

1. Prepare a short research report on the chemical reaction that takes place during oxyfuel gas cutting. Gather information from the library, a chemistry teacher, or people in the welding industry.

2. Secure literature from manufacturers of welding equipment, and study the sections on welding tips. Prepare a list of special-purpose cutting tips and the purposes for which they were designed. Example: Rivet-cutting tip—for removal of rivets.

3. Examine the cutting torches in your shop to determine if they are complete units or cutting attachments. Compare the two types of cutting torches and locate their valves.

4. Write to manufacturers of protective equipment for welding to ask for literature. Prepare a short report on the recommended lens shades (depth of color) for goggles used when flame cutting.

Unit 22

CUTTING PROCESSES

Cutting Skills

Cutting steel and other ferrous metals with an oxyfuel gas cutting torch requires considerable practice to produce clean and accurate edges. The skilled welder can cut straight lines, curves, bevels, and extremely thick steel with precision, **Figure 22-1.** In this unit, you will become familiar with the assembly and operation of an oxyfuel gas cutting outfit and procedures affecting the quality of the cut.

Preparing the Equipment and Material

Before beginning the process of flame cutting thick steel, the following preparations must be made:

1. Study the practice plan, **Figure 22-2.**

2. Obtain the necessary material. Use soapstone to draw the cutting guide lines.

3. Assemble the oxyfuel gas cutting outfit in the same way the oxyfuel gas welding outfit was assembled (see Unit 8).

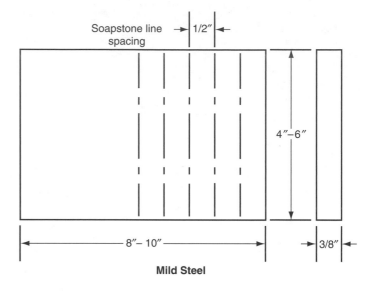

Figure 22-2. Plan for flame cutting mild steel.

4. Attach flashback arrestors and a cutting torch to the hoses or a cutting torch attachment to a torch body that is already properly connected to the hoses.

5. Consult the manufacturer's chart to determine the correct tip size for the metal thickness to be cut. Using a larger tip than necessary wastes oxygen and produces a wide kerf (slot) with uneven edges.

6. Use a clean tip. Dirt or slag buildup can cause a ragged flame. For this activity, use a tip cleaner one drill size smaller than the oxygen or preheat orifices. Lightly file the end of the tip if necessary.

7. Set the correct pressures on the regulators so a neutral flame can be properly adjusted. Improper flame adjustment will cause a rough cut. Be sure to maintain proper oxygen pressure. If the oxygen pressure is too high or too low, poor quality cutting will result. Consult the manufacturer's chart for the recommended oxygen pressure setting.

8. Know the proper cutting speed for the thickness of the metal being cut. Manufacturers of cutting torches and tips provide charts with recommended travel speed. Maintaining the correct travel speed is a skill obtained through practice.

Figure 22-1. A skilled welder uses a cutting torch to make an accurate cut on 10" (254 mm) thick steel. (Smith Welding Equipment Co.)

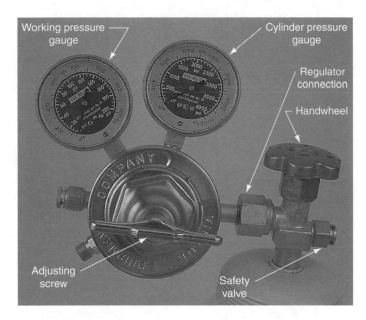

Figure 22-3. A heavy-duty oxygen regulator is needed to safely handle the higher operating pressure of the cutting oxygen. (Victor Equipment Co.)

Procedures for Cutting Thick Steel

The following procedures are used with a one-piece cutting torch (*not* a cutting torch attachment):

1. Set the oxygen regulator to 25 psig (172 kPa) and the acetylene regulator to 5 psig (34 kPa). When cutting extremely thick steel, a heavy-duty oxygen regulator is needed to handle the high operating pressure of oxygen, **Figure 22-3.** If using MAPP gas, set the fuel gas regulator to 5 psig (34 kPa).

2. Open the acetylene torch valve approximately 1/16 of a turn. Hold the tip of the cutting torch facing down. While holding the spark lighter 1″ (25 mm) from the tip, light the preheating flames. If any of the preheating orifices do not light, or if they burn unevenly, shut off the torch and clean the orifices with a tip cleaner.

3. Open the oxygen torch valve and adjust to a neutral flame, **Figure 22-4.** Press the cutting oxygen lever and observe the preheating flames. Compare the cutting flame *without* the cutting oxygen to the cutting flame *with* the cutting oxygen, **Figure 22-5.** The preheating flames should not be affected; they should remain neutral. If the preheating flames grow longer (they turn from neutral to carburizing), adjust the oxygen torch valve so a neutral flame is maintained. The preheating cones should burn with a neutral flame when the cutting oxygen lever is completely open or closed.

4. Position the steel plate on a cutting table, **Figure 22-6.**

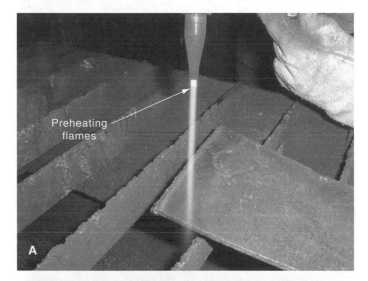

Figure 22-5. Cutting flames. A—Without cutting oxygen. B.—With cutting oxygen.

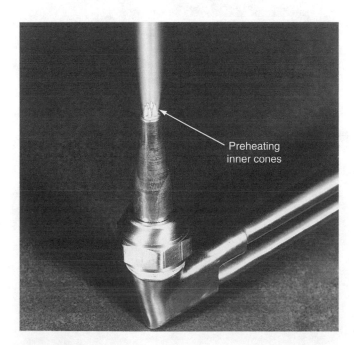

Figure 22-4. Proper adjustment of a neutral preheating flame. (Victor Equipment Co.)

Figure 22-6. A steel grate cutting table allows slag and discarded steel pieces to fall away. Notice protective clothing is being worn.

Figure 22-8. The cutting process begins when the oxygen lever is pressed to release the heavy stream of cutting oxygen.

the cutting action stops, release the oxygen cutting lever and preheat the area at the end of the kerf. Continue the cut when the steel turns cherry red.

7. Practice cutting straight lines until you can make them to acceptable standards. If a high degree of accuracy is desired, clamp a piece of angle iron across the plate to serve as a guide for the torch. Circles and irregular shapes can be cut by hand following a soapstone line. Soapstone works better than chalk because the line is not blown away

5. Hold the torch tip at a right angle to the plate with the inner cones approximately 1/8″ (3.2 mm) above the work. Heat the edge of the plate at one of your guidelines to a cherry red color, **Figure 22-7.** Next, press the oxygen cutting lever to begin cutting. Move the torch slowly forward to continue cutting. **Note:** If MAPP gas is used in place of acetylene, the inner cones of the preheating flame should be 1/4″ to 3/8″ (6.4 mm to 9.6 mm) from the workpiece.

6. Torch movement should be just fast enough to maintain a smooth, even cut, **Figure 22-8.** Flame cutting must be done with two hands. One technique is to grasp the torch handle firmly with one hand while the other hand gently supports the torch tubes. Push or pull the torch smoothly through the palm of the supporting hand, **Figure 22-9.** If

Figure 22-7. Steel plate is being preheated to cherry red before cutting begins. Note the welder's right index finger has not yet pressed the cutting oxygen lever.

Figure 22-9. A steady motion guides this continuous cut by hand.

by the cutting action. Very accurate circles can be cut using a circle cutting attachment fastened to the torch tip, **Figure 22-10.** Making straight cuts is easier with a free-wheel cutting attachment, **Figure 22-11.**

Important: When using a cutting torch attachment, the procedure has an additional step. In Step 3, the cutting oxygen valve must be opened one full turn, which will allow the cutting oxygen to flow once the cutting oxygen lever is pressed.

Shutting Down the Cutting Outfit

After the cutting operation is finished, or whenever the cutting torch operator must leave the work area, the oxyfuel gas cutting outfit must be shut down. Follow these procedures for shutting down the equipment:

1. Shut off the acetylene torch valve.

2. Shut off the oxygen torch valve.

3. Shut off the acetylene cylinder valve.

4. Shut off the oxygen cylinder valve.

5. Reopen the oxygen and acetylene torch valves to release any gas still remaining in the hoses.

6. Once the regulator gauges read zero, shut off the oxygen and acetylene torch valves.

7. Turn the oxygen and acetylene regulator adjusting screws counterclockwise until no pressure remains.

Important: When using a cutting torch attachment, the shutdown procedure has an extra step. The cutting oxygen valve should be closed after Step 7.

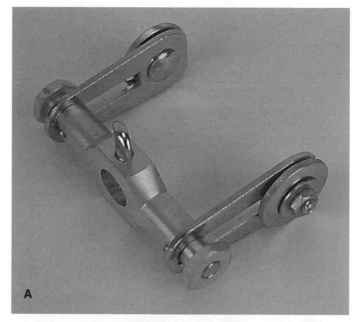

A

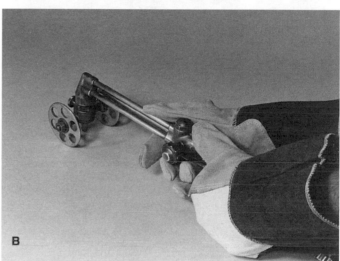

B

Figure 22-11. The cutting torch is easily connected to a straight line cutting attachment. A—Straight line cutting attachment. B—The attachment is designed to hold the torch tip steady for the length of the cut. It can be used freehand or with a straightedge.

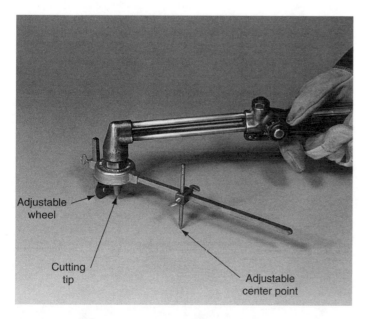

Adjustable wheel

Cutting tip

Adjustable center point

Figure 22-10. Setup for a circle cutting attachment.

Check Your Progress

Write your answers in the spaces provided.

1. Manufacturers of cutting torches and tips provide charts with information regarding:

    a. _____

    b. _____

    c. _____

2. Three factors that may affect the quality of a cut are:

    a. _____

    b. _____

    c. _____

3. When using acetylene gas, hold the torch tip at a _____ to the plate with the inner cones approximately _____ above the work.

4. The preheating cones should burn with a neutral flame when the _____ is completely open or closed.

5. Why is an extra step needed when lighting a cutting torch attachment as opposed to lighting a one-piece cutting torch? _____

_____

_____

Things to Do

1. Inspect the cutting tips in your shop and clean any dirty ones. Use the correct size tip cleaners and a smooth tip cleaning file.

2. With soapstone, lay out a 3″ (7.6 cm) diameter circle on a scrap piece of 1/4″ (6.3 mm) thick mild steel plate. Follow the line and try to cut out the circular disc as accurately as possible.

3. In the space below, make a sketch of one of your best cuts on 3/8″ (9.5 mm) steel plate. Show the texture of the cut edge, amount of slag on the bottom edge, and any melting of the top edge of the plate.

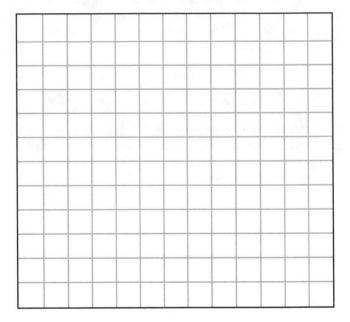

4. After you have become fairly skilled in cutting plate, make practice cuts on scrap steel angle iron, round bar stock, and pipe.

5. Visit a local welding or metal fabricating company, and observe the types of oxyfuel gas cutting the welders are required to perform. Obtain samples of scrap pieces that have been cut and show them to your class.

Unit 23

LAP JOINT WELDING THICK STEEL

Thick Steel Modifications

Lap joint welds on thick steel plate are very similar to those made on thin steel. The torch and rod motion is the same as the weaving motion used to run beads with welding rod on thin steel. Making lap joints on thick steel, as opposed to thin steel, requires several changes to the welding process. They include:

- A much larger molten weld pool.
- More filler metal.
- More torch and welding rod movement because the weld bead is much wider.
- More heat to melt the base metal and form the weld pool, which requires a larger welding torch tip.

Using more heat increases the possibility of undercut; however, the thick edge of the top plate will not melt as quickly. The alternating weaving motion between the edge of the top plate and the surface of the bottom plate helps eliminate undercut, since the heat does not stay directed at the top edge.

Note: Beginning welders should be careful not to move the torch and welding rod too quickly. A rapid weaving motion can produce very weak welds with little or no penetration, even though the weld bead looks good on the surface.

Preparing the Equipment and Material

Before beginning the process of lap welding thick steel, the following preparations must be made:

1. Study the practice plan, **Figure 23-1.**

2. Cut the material to size, and remove any burrs and sharp edges.

3. Make sure the plates are flat and straight. Make a trial assembly to also ensure they are in close contact. If they are not, flatten the plates with a hammer.

4. Obtain 1/8″ or 3/16″ (3.2 mm or 4.8 mm) diameter mild steel welding rod.

5. Select the correct tip size according to the manufacturer's specifications for your equipment.

Procedures for Welding Lap Joints on Thick Steel

Follow these procedures for lap joint welding thick steel:

1. Position the plates on the welding bench. Overlap them as specified and tack weld each end.

2. Set both the oxygen and acetylene pressure regulators to 8 psig (55 kPa). Pressure is increased slightly when welding thicker steel. Light the torch and adjust it to a neutral flame.

3. Begin the weld by forming a molten pool at the right edge. Add filler rod and start moving the pool along the joint.

4. Move the torch and rod in a weaving motion, as previously described. Keep the bottom of the pool slightly ahead of the top.

5. Add enough filler rod to form a slightly convex (outward) arc in the bead, **Figure 23-2.**

6. Complete the weld and examine it for uniformity, smoothness, and penetration, **Figure 23-3.**

7. Continue to practice lap welds on thick steel until you can consistently make them to acceptable standards. Practice welds should be made on both sides of the plates to conserve metal.

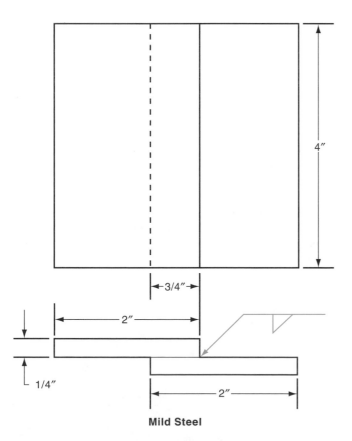

Mild Steel

Figure 23-1. Plan for a lap weld on thick steel plate.

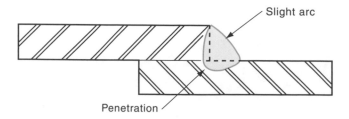

Figure 23-2. A good lap weld will have a slightly convex arc on the bead and even penetration into both plates.

8. After achieving quality lap joint welds in the horizontal position, practice making them on thick steel in the vertical and overhead positions.

Check Your Progress

Write your answers in the spaces provided.

1. When lap welding thick steel rather than thin steel, modifications to the welding process include:

   a. _____

   b. _____

   c. _____

   d. _____

2. Why is undercut a greater problem when welding on heavy plate? _____

   _____

3. How is it possible to create a very weak lap joint weld that looks good? _____

   _____

4. The steel plates should be _____, _____, and _____.

5. When making lap joint welds on thick steel, add enough filler rod to form a slightly _____ arc in the bead.

Instructor's Initials _____ Date _____

Things to Do

1. Refer to the plan in Figure 23-1 and make a setup for lap welding. This time use the backhand method. In the space provided, describe what was different about welding backhand compared to forehand. _____

   _____

   _____

   _____

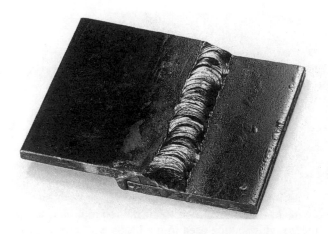

Figure 23-3. A completed lap weld on thick steel plate.

2. In the space provided, make a sketch of one of your best lap welds and indicate any imperfections in the weld.

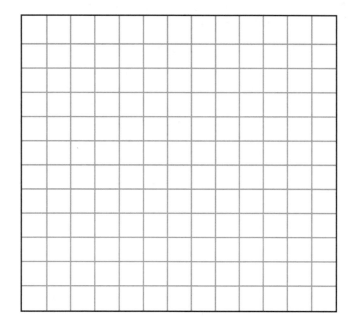

3. Visit a metalworking or woodworking shop in your building or at a local school. Compile a list of machine parts or equipment made with lap welds. In each case, consider why a lap weld joint was selected over another type of joint.

Unit 24

T-JOINT WELDING THICK STEEL

Thick Steel Modifications

The technique for welding T-joints on thick steel plate is similar to welding T-joints on thin steel. Modifications to the welding process for lap joints on thick steel also apply to fillet welds on T-joints, with two additional requirements:

- Direct the torch at a 45° work angle, splitting the angle between the two plates, **Figure 24-1.** Use a 75° travel angle.
- Move the torch and rod in a weaving motion to help prevent undercutting of the vertical plate.

Preparing the Equipment and Material

Before beginning the process of T-joint welding thick steel, the following preparations must be made:

1. Study the practice plan, **Figure 24-2.**

2. Cut the material to size and remove all sharp edges.

3. Obtain 1/8″ or 3/16″ (3.2 mm or 4.8 mm) diameter mild steel welding rod.

4. Select the appropriate welding torch tip from the manufacturer's reference chart.

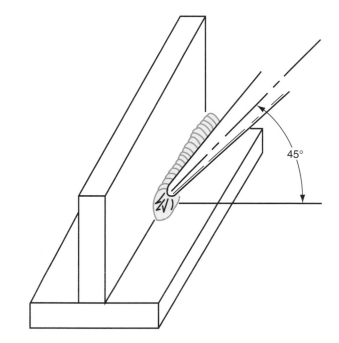

Figure 24-1. A 45° work angle and a 75° travel angle are used to make fillet welds on thick T-joints.

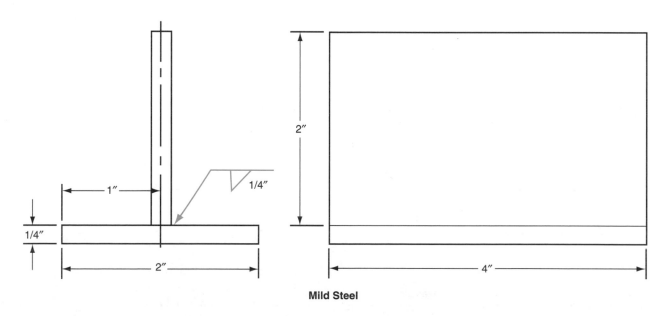

Mild Steel

Figure 24-2. Plan for making a fillet weld on a thick steel T-joint.

Procedures for Welding T-Joints on Thick Steel

Follow these procedures for T-joint welding thick steel:

1. Position the material on the welding table so the vertical plate is in the center of the flat plate and at a 90° angle to the surface.

2. Using a neutral flame, tack weld the ends of the plates.

3. Form a molten pool at the right-hand corner. Add welding rod and begin welding. Keep the tip of the flame centered in the corner of the joint. Take care not to undercut the vertical plate.

4. Keep the edge of the molten pool on the flat plate ahead of the part on the vertical plate, **Figure 24-3.** Add filler rod to the upper part of the pool. Practice welding with either a weaving or semicircular torch motion.

5. Bend or break the weld as you did when testing the fillet weld on thin steel sheet. Inspect the weld for penetration, fusion, porosity, even bead width, uniform ripples, and undercutting.

6. After the weld has cooled, check the final angle between the vertical and horizontal plates. The angle is often less than 90° due to shrinkage of the fillet weld during cooling. To prevent shrinkage, practice tack welding some pieces at angles between 91° and 95°.

7. Continue to practice T-joint welds on thick steel until you can consistently make them to acceptable standards, including a final angle of 90°.

8. After achieving quality T-joint welds in the horizontal position, practice making them on thick steel in the vertical and overhead positions.

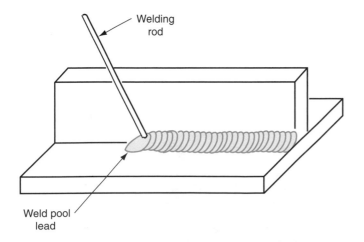

Figure 24-3. The weld pool on the flat plate should lead the molten part on the vertical plate. Welding rod is added to the upper part of the pool.

Check Your Progress

Write your answers in the spaces provided.

1. Modifications to the welding process for lap joints on thick steel also apply to _____ on T-joints.

2. Two additional requirements for making a fillet weld on a T-joint are:

   a. _____

   _____

   b. _____

   _____

3. Where must the edge of the molten pool be kept during welding?_____

   _____

   _____

4. What can be done to allow for shrinkage of heated metal plates so the final angle between the vertical and horizontal plates is 90°?_____

   _____

   _____

5. When inspecting the fillet weld on a T-joint, check for:

   a. _____

   b. _____

   c. _____

   d. _____

   e. _____

   f. _____

Instructor's Initials_____ Date _____

Things to Do

1. Cut more pieces to size and position the plates for T-joint welding. Weld the plates using the backhand method. In the space provided, describe what was different about welding backhand compared to forehand._____

   _____

   _____

   _____

   _____

2. Using your shop library and other reference materials, prepare a written report on the various applications of T-joint welding to industrial products. You may want to visit a local welding fabrication company for additional information to include in your report.

3. On the joints shown, sketch the weld beads for each T-joint as indicated by the welding symbol. Be sure to show penetration of the bead into each plate.

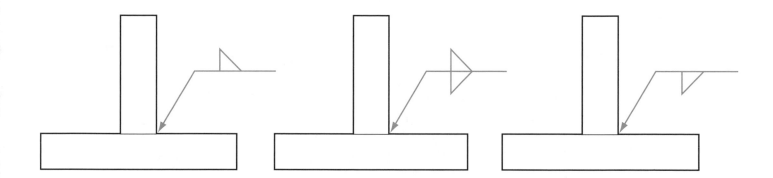

Unit 25

BUTT JOINT WELDING THICK STEEL

Joint Preparation

Thick steel generally is 3/16″ (4.8 mm) or greater. Producing strong butt weld joints on thick steel requires careful joint preparation. The quality of the weld depends on edge preparation, alignment of the plates, bead width, and penetration.

The edges of thick steel must be specially prepared as a V-groove, J-groove, U-groove, or bevel-groove, **Figure 25-1.** Butt welds on thick steel can be done on one or both sides, depending on the thickness of the metal. Welding on both sides using double grooves is recommended on metal over 1/2″ (12.7 mm). Two-sided welding ensures the strongest weld possible by providing complete penetration.

Edges of thick steel are machined, ground, or flame cut to produce the appropriate groove. The single square-groove weld can be used on thin steel up to 3/16″ (4.8 mm) and does not require special joint preparation. See **Figure 25-2.**

Butt welding on thick steel requires more than one weld bead, whereas thin steel requires only one bead. One weld pass is needed to produce each weld bead. Each weld bead should be 1/4″ to 3/8″ (6.4 mm to 9.6 mm) wide. Wide beads are usually weak because impurities and gases are trapped inside when the beads cool.

Preparing the Equipment and Material

Before beginning the process of butt joint welding thick steel, the following preparations must be made:

1. Study the practice plan, **Figure 25-3.**

2. Cut the material to size, and remove any burrs and sharp edges.

3. On a coarse grinding wheel, grind a 30° bevel on each plate, **Figure 25-4.** A cutting torch can also be used. Remove any slag if bevels are flame cut.

4. Obtain 1/8″ or 3/16″ (3.2 mm or 4.8 mm) mild steel welding rod.

5. Select the correct tip size according to the manufacturer's specifications for your equipment.

Procedures for Welding Butt Joints on Thick Steel

Follow these procedures for butt joint welding thick steel:

1. Align the plates on top of firebricks to form a butt joint. Use 1/8″ (3.2 mm) diameter, U-shaped welding rod to space the plates with a straight, even 1/8″ (3.2 mm) root opening.

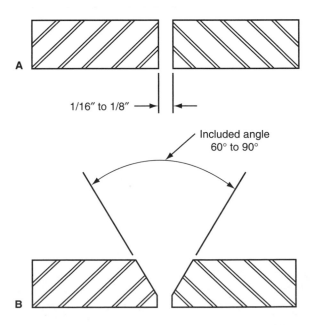

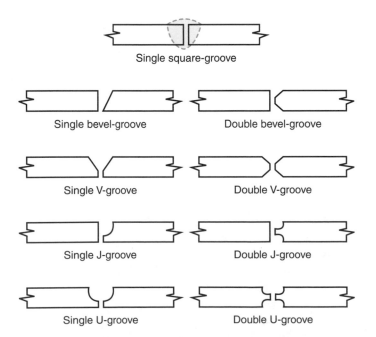

Figure 25-1. Butt joint edges are prepared in a variety of shapes. Double grooves are welded from both sides for complete penetration.

Single square-groove

Single bevel-groove   Double bevel-groove

Single V-groove   Double V-groove

Single J-groove   Double J-groove

Single U-groove   Double U-groove

Figure 25-2. Joint preparation. A—Single square-groove butt joint used on thin steel. B—Single V-groove butt joint used on thick steel.

1/16″ to 1/8″

Included angle 60° to 90°

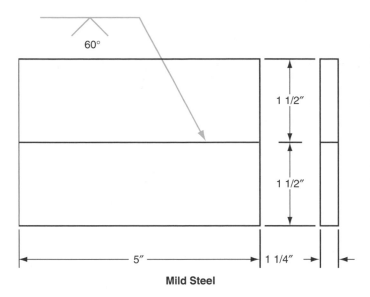

Figure 25-3. Plan for a butt welding joint.

2. Adjust the torch to a neutral flame. Place 3/8″ to 1/2″ (9.6 mm to 12.7 mm) long tack welds at each end of the plate.

3. Deposit the first weld bead from the right side. Use a side-to-side torch motion to melt the edges of both plates. The torch should be held at a 90° work angle and a 35° to 40° travel angle.

4. To maintain a straight weld bead, center the torch and welding rod over the root opening, **Figure 25-5.** Melt the edges of each plate evenly for complete penetration. Approximately 1/16″ (1.6 mm) of each edge of the joint should be melted away. A molten keyhole with a diameter of approximately 3/16″ (4.8 mm) will form. Maintain the keyhole shape the entire length of the weld.

5. Add the welding rod at the center edge of the keyhole. Slowly move the torch along the joint and create a weld slightly over 1/4″ (6.4 mm) wide.

6. Check your first weld bead. The weld face should be slightly concave and fill 2/3 to 3/4 of the depth of the V-groove. Check the root penetration to verify a 1/16″ (1.6 mm) convex bead has been formed on the bottom side of the plate.

Figure 25-5. The tack-welded plates are aligned over the firebricks. The torch tip and filler rod are directly over the center line of the butt joint.

7. Deposit the second weld bead using the same torch angles; however, reverse the travel direction by using the back-hand method. Use an alternating torch and rod motion in which the torch is at one edge of the weld pool when the welding rod is at the opposite edge.

Note: A larger amount of welding rod must be added to form the second weld bead since the gap to be filled is wider. The welding rod should be in contact with the weld pool at all times. Dipping action is not necessary.

8. Check your second weld bead. Complete penetration into the sides of the joint and the first weld bead is required. The second weld bead should be slightly convex.

9. Practice butt welding on thick steel until your joints meet acceptable standards, **Figure 25-6.**

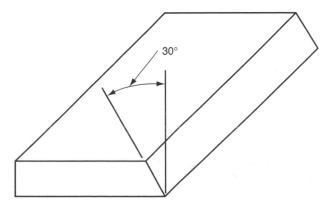

Figure 25-4. A bevel is ground on each plate at 30° to the vertical.

Figure 25-6. Completed single V-groove butt weld. Notice the even bead ripple and straightness.

10. To test your welds, cut the plate into 1″ (25.4 mm) wide strips across the weld. Grind down the face and root of the weld bead until they are flush with the rest of the plate. Place the strip in a guided bend jig with the face of the weld down. Bend the sample strip until it breaks or a U-shape is formed. Inspect for defects. A good weld will bend into a U without showing any signs of cracking, **Figure 25-7.**

11. Repeat the test procedure, but place the next strip in the jig with the root of the weld facing down to inspect the root side. Before grinding and bending, mark each side of the strips with an "F" for face side or an "R" for root side.

Check Your Progress

Write your answers in the spaces provided.

1. The quality of a butt weld depends on:

   a. _____

   b. _____

   c. _____

   d. _____

2. Butt welds on thick steel require specially prepared joints. Four common types of butt-welded joints are:

   a. _____

   b. _____

   c. _____

   d. _____

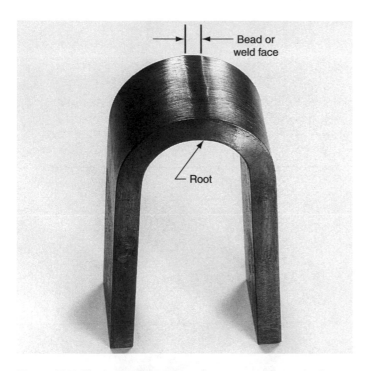

Figure 25-7. The butt weld test piece shows no cracking at the face or edge of weld.

3. Complete penetration can be achieved with a single square-groove weld on thin steel up to _____.

4. Why is two-sided welding using double grooves recommended on metal over 1/2″ (12.7 mm)? _____

   _____

   _____

5. Why are wide beads usually weak?_____

   _____

   _____

Instructor's Initials _____ Date _____

Things to Do

1. Vary the angle of the bevel for a single V-groove butt weld. Practice making welds that meet acceptable standards.

2. Visit a local welding fabrication company. Prepare a list of the products they make using butt welds on thick steel plate.

3. Draw in the welding symbols for the butt welds shown.

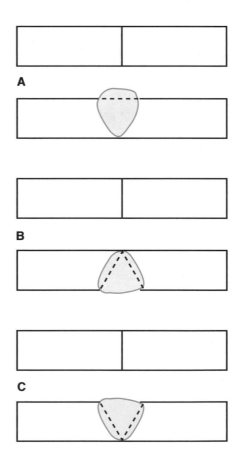

4. Using library or shop reference materials, write a report on butt welding heavy steel plate of 1/2″ (12.7 mm) thickness or greater. Discuss type of joint preparation, rod sizes, and torch and rod motion.

Unit 26

WELDING ALUMINUM

Aluminum Alloys

Aluminum refers to an entire family of metals. Pure aluminum is a nonferrous metallic chemical element. In practice, it is used as an alloy. The aluminum family ranges from very soft, commercially pure aluminum (99.5%) to high-alloy aluminum that contains magnesium, silicon, manganese, or copper to increase its strength and other properties.

Aluminum alloys are classified by a three- or four-digit identification system. Three-digit designations are used for cast aluminum. Four-digit designations are used for wrought aluminum, **Figure 26-1.** Wrought aluminum includes sheet, plate, extrusions, pipe, and other forms that are not cast.

Oxyacetylene welding is not the preferred process for welding aluminum. However, oxyacetylene welding aluminum alloys can produce satisfactory results. An aluminum alloy filler rod is used for most joint work and a quality flux is required. Most joint design principles for welding steel apply to aluminum. The lap joint is not recommended because fluxes may become trapped between the surfaces and cause corrosion. A backing is often used for certain joints to obtain complete penetration without burning through, **Figure 26-2.**

Characteristics of Aluminum

Aluminum has certain characteristics the welder should keep in mind:
- Aluminum is one of a number of nonferrous metals with hot shortness. *Hot shortness* means a metal becomes very weak when hot. The strength of the hot metal cannot support its own weight, causing the heated area to fall

through the metal and leave a large hole. To prevent this from happening, the metal should be adequately supported during welding.
- Aluminum has an exact melting point. The light, silvery metal does not change color before melting, so no indication is given of when it will suddenly melt and collapse.
- Since molten aluminum oxidizes very quickly, a good flux must be used to prevent a heavy coating of oxides from forming on the surface.
- Aluminum is an excellent heat conductor. Preheating aluminum over 3/8″ (9.5 mm) thick is recommended to avoid stresses and to more easily maintain the weld pool.

Preparing the Equipment and Material

Before beginning the process of welding aluminum, the following preparations must be made:

1. Study the practice plan, **Figure 26-3.**

2. Cut the material to size.

3. Prepare a V-joint with a 90° included angle.

4. Obtain 3/16″ (4.8 mm) diameter welding rod and paste flux. Welding rod should be appropriate for the aluminum alloy being welded. Follow the manufacturer's specifications for application and removal of the flux.

5. Select a torch tip one size larger than is used for welding the same thickness of steel.

Procedures for Welding Aluminum

When welding aluminum, an oxyacetylene flame should be used because it produces the highest temperature. Follow these procedures:

1. Thoroughly clean the joint area with a stainless steel wire brush.

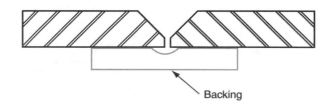

Figure 26-2. Burnthrough can be eliminated by using a backing when attempting to gain complete penetration.

Aluminum Association Alloy Group Designation	Major Alloying Element	Example
1XXX	99% aluminum (minimum)	1100
2XXX	copper	2024
3XXX	manganese	3003
4XXX	silicon	4043
5XXX	magnesium	5052
6XXX	magnesium and silicon	6061
7XXX	zinc	7075
8XXX	other	—

Figure 26-1. Classifications of wrought aluminum alloys.

2. Position the plates flat on the fire bricks with 1/8″ (3.2 mm) root opening.

3. Brush a coating of paste flux on the filler metal and along the edges of the V-joint.

4. Adjust the torch to a neutral or slightly reducing (carburizing) flame.

5. Using a small circular motion, place the flame over the right-hand edge of the joint until the flux melts. Rub the rod over the heated area every few seconds until you feel the aluminum base metal soften at the touch of the filler metal. Tack weld both ends of the joint.

6. Repeat the process of creating a molten pool. When the aluminum base metal becomes soft, keep the filler metal in the softened weld pool and begin welding. Caution: The exact welding temperature is critical! Excess heat will melt a hole in the joint due to the hot shortness of aluminum. With too little heat, the metals will not fuse.

7. Hold the torch at a 90° work angle and a 60° travel angle. Use the forehand welding technique. Adjust the angle as needed to maintain an even-flowing pool. Little torch motion is required other than what is necessary to carry the pool forward.

8. Deposit enough filler metal to form a bead that completely fills the joint. Try to maintain a straight, uniform bead as the weld progresses.

9. The completed weld should be cleaned as soon as possible. Any flux left on an aluminum weld will soon cause corrosion. Scrub the weld with a fiber brush in very hot water; then rinse and dry it. **Figure 26-4** shows a completed butt weld.

10. Examine the weld bead for straightness, roundness, and clean appearance. Cut the plate across the bead and inspect for even penetration and good fusion.

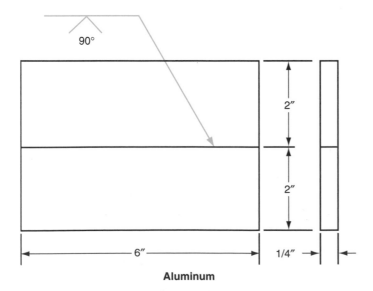

Figure 26-3. Plan for welding aluminum.

Figure 26-4. A properly made weld on aluminum.

Check Your Progress

Write your answers in the spaces provided.

1. How much aluminum does commercially pure aluminum contain *(circle letter)*?
   a. 92.5%
   b. 96.4%
   c. 99.5%
   d. 99.9%

2. Two classifications used to describe aluminum material are:

   a. _____

   b. _____

3. Hot shortness means *(circle letter)*:
   a. metal gets shorter when heated.
   b. metal loses its strength at high temperatures.
   c. shorter metal requires additional preheat before starting to weld.
   d. flux is used to add strength to a weld joint at high temperatures.

4. What should be done to prevent a molten weld pool from falling through the metal workpiece? _____

   _____

5. Why is the lap joint *not* recommended for welding aluminum?_____

   _____

   _____

6. Aluminum is an excellent _____.
   Thus, preheating a part may be necessary.

7. Compared to a torch tip for welding steel, what tip size should be used for welding aluminum? _____

   _____

8. What type of brush should be used to clean aluminum before welding?_____

   _____

9. What type of flame should be used when welding aluminum? _____

_____

10. Why should an aluminum weld be cleaned as soon as possible? _____

_____

Instructor's Initials _____ Date _____

Things to Do

1. Research the various aluminum alloys weldable with an oxyacetylene torch. Prepare a list of the alloys, and include the manufacturers' recommendations for welding rods and fluxes to be used with each alloy.

2. Obtain samples of as many different types of aluminum welds as possible. Display the samples on a bulletin board. Indicate the use or purpose of each weld in the display.

3. Use your shop library and other reference materials to prepare a report on other processes used to weld aluminum.

4. Practice welding aluminum with different joint designs in both sheet and thick plate. Ask your instructor to check your welds.

5. Obtain a broken aluminum casting. Prepare the edges and repair the piece with oxyacetylene welding.

Unit 27

BRAZING PROCESSES

Two Metal-joining Processes

Brazing is a metal-joining process. It involves the use of a filler metal that has a liquidus above 840°F (450°C) but below the solidus of the metals or alloys to be joined. *Liquidus* is the temperature at which a metal or alloy is completely liquid. *Solidus* is the temperature at which a metal begins to melt. Brazing filler metals are nonferrous alloys.

Brazing and braze welding are two similar metal-joining processes with some distinctions. In *brazing,* the metals to be joined are fitted tightly together over a large surface area. A lap joint, which creates a large surface area, is typical, **Figure 27-1.** The mating surfaces must be properly aligned and fitted. The molten filler metal enters and fills the joint by capillary action.

Capillary action is the process in which a liquid is drawn into the space between two tightly fitted surfaces.

The strength of the brazed joint depends upon molecular attraction and penetration of the filler metal into the surface of the base metal. Molecular attraction occurs when the smallest particles of each metal combine and form a bond. To obtain the greatest joint strength, the metal parts must be closely fitted, and a thin uniform layer of filler metal must flow between the surfaces. A brazed joint is as strong as the bond between the filler metal and base metal.

Like brazing, *braze welding* uses a filler metal with a liquidus above 840°F (450°C). However, when braze welding, the joint is filled by running a braze pool to fill a groove joint or form a fillet. The filler metal does not fill the joint by capillary action. Joints that can be braze welded include lap, corner, T, edge, and butt. The strength of the joint is determined by the strength of the filler metal.

Brazing and braze welding are very important industrial processes. They are particularly valuable in joining or repairing mild steel, cast iron, copper, and brass. These processes are also used to join dissimilar (unlike) metals, such as cast iron and brass. Brazing is used extensively in joining thin steel sheet, fastening carbide cutting tips to steel tools, and carrying out a variety of other applications. The following units discuss brazing and braze welding steel using an oxyfuel flame and a copper-zinc alloy filler material (called bronze rod).

Advantages of Brazing Processes

Lower temperatures used in brazing and braze welding offer several advantages:

- The base metal is not melted. Thus, a material such as malleable cast iron can be joined without losing its ductility (ability to be stretched).
- Stresses that develop in the fusion welding of certain metals are minimized or eliminated. Less expansion and contraction occur as well.
- Preheating and postheating of cast iron is reduced, minimizing cracks which easily occur during fusion welding when cast iron is heated or cooled too quickly.
- Braze welding is used for building up metal surfaces using soft materials, such as brass, bronze, or other alloys. Less heat is generated, and the soft metal can be easily smoothed by filing or power sanding.

Filler Metals for Brazing

Brazing filler metals have a liquidus temperature above 840°F (450°C). Each filler metal alloy is designated by the

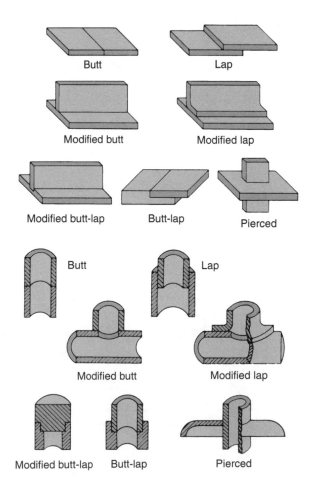

Figure 27-1. Well-designed joints for brazing. (Handy & Harman/Lucas-Milhaupt, Inc.)

Base Metals	Al and Al Alloys	Cu and Cu Alloys	Carbon and Low-alloy Steels	Cast Iron	Stainless Steels
Al and Al alloys	BAlSi				
Cu and Cu alloys	Combination not recommended	BAg, BAu, BCuP , RBCuZn			
Carbon and low-alloy steels	BAlSi	BAg, BAu, RBCuZn, BNi	BAg, BAu, BCu, RBCuZn, BNi		
Cast iron	Combination not recommended	BAg, BAu, RBCuZn, BNi	BAg, RBCuZn, BNi	BAg, RBCuZn, BNi	
Stainless steels	BAlSi	BAg, BAu	BAg, BAu	BAg, BAu, BCu, BNi	BAg, BAu, BCu, BNi

Key:
BAlSi	Aluminum-silicon	BCu	Copper base
BAg	Silver base	RBCuZn	Copper-zinc
BAu	Gold base	BNi	Nickel base
BCuP	Copper-phosphorus		

Figure 27-2. Shaded boxes show recommended brazing filler metals for common base metal combinations.

letter B (for brazing alloy) and the chemical symbols of the major elements in the alloy. See **Figure 27-2.**

A number specifies the type of alloy, for example, BAg-1, BAg-2, BAg-3. Each of these is a variation of a silver brazing alloy. Filler metal usually comes in wire form in various diameters. Silver and gold brazing alloys are very thin, usually less than 1/64″ (0.4 mm). Aluminum and copper-based filler metals come in diameters from 1/32″ to 1/8″ (0.8 mm to 3.2 mm).

Fluxes for Brazing

Paint, rust, grease, or other foreign matter must be removed from the surfaces to be joined before brazing begins. Cleaning can be done using a file, grinder, aluminum oxide paper, or emery cloth. During brazing, a flux is used to remove oxides and other undesirable substances from the brazing area. A flux also works to prevent additional oxides from forming. A flux acts on both the filler alloy and the metals being joined to keep them clean. Brazing fluxes contain different cleaning agents. The most common compounds include chlorides, fluorides, borates, and boric acid. Fluxes are available in powder, paste, and liquid forms. Brazing filler metals with a flux coating are also available.

When selecting a flux for the brazing process, it is important to consider the type of base metal(s) being joined, the filler metal being used, and the method for removing the flux. Follow the manufacturer's specifications for the proper flux for each brazing job.

Without a flux, the molten brazing filler metal will form into balls and roll across the metal surface without sticking. The image is similar to drops of water on a well-waxed car. Not only must the base metal be clean, it must be free of oxides. Because oxides form quickly on a clean metal surface, they must be removed using a high-quality flux so the filler metal alloy can make a good bond.

Flux is usually applied with a brush to the surfaces to be brazed. In braze welding, flux is applied to the filler metal, which transfers it to the braze weld joint. The heated filler metal is dipped into a powdered flux. The flux sticks to the metal and transfers to the joint when enough heat is supplied by the torch. Paste fluxes may be applied to the filler metal with a small, inexpensive brush. Excess flux must be removed from the base metal surfaces after brazing. Warm-water washing will remove most fluxes. Some scrubbing with a brush may be needed.

Basic Brazing Procedures

The brazing process is quite different from the fusion welding process. You are already familiar with how the weld pool is controlled when welding. In brazing, no pool is formed. Instead, the base metal, with flux applied, is heated to the proper temperature. Filler metal and flux are added and flow into the joint by capillary action. The heat from the base metal melts the filler metal.

Braze welding steel is similar to welding steel except the base metal is not melted. Only the filler metal is melted when braze welding. See **Figure 27-3.** Steel is heated with a torch until its color turns dull red. The filler metal coated with flux is brought in contact with the base metal. The heat from the base metal melts the filler metal. Filler metal is added until the desired bead shape is obtained. The molten metal cannot be easily directed by the flame, as it can in fusion welding. The molten metal tends to run away from the flame but flows back into the heated area as the torch is moved away.

Brazing Safety

Brazing should be done in a properly ventilated area. The fumes of some brazing alloys, such as zinc oxide, are very toxic. Breathing these fumes can cause chills and an upset stomach. Although the illness is not considered serious, the symptoms can be very painful. A nurse or doctor should be consulted immediately.

Some brazing alloys contain cadmium, a very toxic substance. It is best to use a brazing alloy that does not contain cadmium. If an alloy containing cadmium must be used, braze in a well-ventilated area and wear an air-supplied purifier or respirator.

Figure 27-3. Braze welding a T-joint in the flat position.

Observe the same safety precautions for brazing as you would for fusion welding. Wear protective clothing, gloves, and goggles and follow instructions for the proper and safe handling of equipment.

Check Your Progress

Write your answers in the spaces provided.

1. Brazing is a joining process involving the use of a filler metal that has a(n) _____ above 840°F (450°C) but below the _____ of the metals to be joined.

2. Describe how brazing is different from braze welding.

   _____

   _____

   _____

   _____

3. Define *capillary action.* _____

   _____

   _____

4. What is the designation for a filler metal whose main components are copper and zinc? _____

5. How is the greatest joint strength obtained in brazing?

   _____

   _____

6. Which statement(s) describe the advantages of lower temperatures used in brazing *(circle letter or letters)*?
   a. The ductile properties of cast iron are not destroyed.
   b. Fusion of the filler material and base metal is faster.
   c. Preheating and postheating of cast iron are reduced.
   d. Alloy filler material is less expensive than steel welding rod.

7. A(n) _____ keeps the filler metal and metals being joined clean during brazing.

8. What will occur if flux is *not* used during brazing?

   _____

   _____

9. Describe how brazing is different from fusion welding.

   _____

   _____

   _____

   _____

10. What safety precautions should be taken when using a brazing alloy that contains cadmium? _____

   _____

   _____

Instructor's Initials _____ Date _____

Things to Do

1. Cut or shear a piece of 16 gage mild steel to 2″ × 6″ (5 cm × 15 cm). Clean one side of the sheet with abrasive paper and position the sheet on the welding table as shown in the illustration. Perform the following experiment and write down your observations where requested.

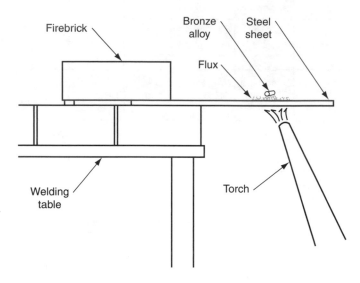

   a. Put a small amount of powdered flux on the clean, topside of the steel sheet.
   b. Cut a 1/8″ (3.2 mm) piece off the end of 1/8″ (3.2 mm) diameter RBCuZn bronze filler metal. Place the piece in the center of the flux.
   c. Adjust the torch to a neutral flame, and heat the bottom side of the plate under the flux and filler metal.
   d. Which melts first, the flux or the filler metal?

   _____

   e. Slowly move the torch around the bottom of the plate after the flux and alloy have melted. Explain how the flux and alloy move. _____

   _____

   _____

   f. What happens if you apply heat until the bronze alloy begins to burn? Note any changes in color on the plate surface. _____

   _____

   _____

   _____

2. The two main purposes of a flux are to keep the metal surface clean and to protect the bead from atmospheric contamination during brazing. Using library and reference materials, write a short report on how a flux performs these functions.

Unit 28

BRAZING LAP JOINTS

Basic Brazing Steps

Brazing is primarily done on lap joints. The brazing joint may be two flat surfaces overlapped, or it may be a pipe or tube joint, **Figure 28-1.** Brazing filler metal is used to fill the gap between the tightly fitted surfaces. Capillary action draws the molten brazing filler metal into the joint.

Overlapping surfaces must be in close contact. Joint strength lies in the bond between the filler metal and the base metals, not in the strength of the filler metal itself. The strongest joint is obtained with minimum spacing between the parts.

Seven basic steps must be followed when making a brazed lap joint:

1. Check the fit of the parts to be joined. Be sure they are flat and have good contact with each other.

2. Clean the surfaces to be brazed.

3. Apply the proper flux to both surfaces to be joined.

4. Properly fit the parts together.

5. Apply heat to the joint. Use care not to overheat the base metal.

6. Add filler metal to fill the joint and obtain complete penetration.

7. Clean the joint after brazing.

Parts to be brazed must be cleaned before a flux is applied. Clean the surfaces to be joined with steel wool, an abrasive cloth, a file, grinder, or chemical cleaning solution. Surface areas that are not clean will prevent the flow of the flux and braze metal. Bare, weak spots will be left in the brazed joint.

Apply a flux over the areas to be brazed. A flux cleans and removes unwanted materials from the metal surfaces. It also prevents the formation of oxides. Brazing filler metal will not flow over an area without a flux coating.

Brazing filler metals are made from various alloys (refer to Figure 27-2). A number of brazing alloys are made with silver. *Silver brazing* describes the brazing process when silver brazing filler metal is used. "Silver soldering" is an improper term sometimes used to describe the silver brazing process.

Silver brazing is widely used for joining both ferrous and nonferrous metals. The process is ideal for joining small precision parts, tubing, and copper and brass fittings. Mounting carbide tips on cutting tools is another important use of silver brazing, **Figure 28-2.** It is also one of the major processes used in the fabrication of jewelry, air conditioning and refrigeration units, bicycles, airplane parts, and electrical and household appliances.

This unit will give you an opportunity to acquire skills in making brazed lap joints on steel sheet. The same techniques can be used with other base metals. The flux and brazing filler metal must be appropriate for the base metal being brazed.

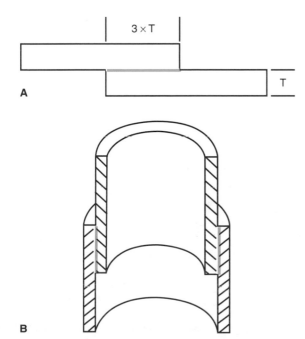

Figure 28-1. Overlap for a brazed lap joint should be three times the metal thickness. A—Lap joint. B—Pipe fitting lap joint.

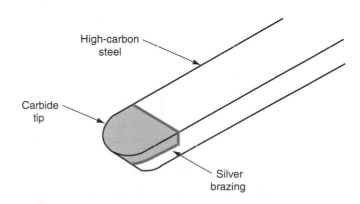

Figure 28-2. A carbide tip silver brazed on a round-nose wood turning lathe provides a long-lasting cutting edge.

Preparing the Equipment and Material

Before beginning the process of lap joint brazing, the following preparations must be made:

1. Study the practice plan, **Figure 28-3.**

2. Cut the material to size, and remove all burrs and sharp edges.

3. Be sure the metal is flat and adjoining surfaces have good contact.

4. Obtain 1/16″ (1.6 mm) BCu or BRCuZn filler metal and the proper flux.

5. Select the correct torch tip size for the metal thickness.

Procedures for Lap Joint Brazing

Follow these procedures for brazing lap joints:

1. Clean the overlapping surfaces and joint edges with a stainless steel wire brush, emery cloth, or grinding wheel. Remove all surface oxides and contaminants. Wipe the surfaces with a clean cloth to remove residual particles from the joint area.

2. Apply flux to both surfaces. Completely cover the 3/8″ (9.6 mm) area to be brazed.

3. Place the metal pieces on firebricks, and position the pieces so they overlap 3/8″ (9.6 mm).

4. Adjust the torch to a neutral flame.

5. Heat the end of the filler metal and dip it into the flux. Flux will stick to the filler metal.

6. Use the flame to heat the center of the overlapped area, **Figure 28-4.** Distribute the heat by moving the torch in a circular motion. Do not direct the flame at the edge of the top piece only. Keep the inner cone away from the metal surface to prevent melting the base metal. The flux will heat and begin to boil, indicating the metal is getting close to brazing temperature.

7. Touch the brazing filler metal to the edge of the joint. When the correct temperature is reached, the heat in the base metal will melt the filler metal. The filler metal will be pulled into the joint by capillary action.

8. Continue moving the flame down the length of the lap joint in a circular motion, adding the filler metal so it is drawn into the joint. Add filler metal to fill the joint without creating a fillet at the edge of the joint.

9. Let the completed joint cool; then examine both edges. If the joint was properly cleaned, fluxed, and heated, a thin line of filler metal should be visible on both sides.

10. Thoroughly wash the joint in warm water to remove excess flux from the surfaces.

Procedures for Silver Brazing

Follow these procedures for silver brazing:

1. Obtain two 1/16″ (1.6 mm) thick pieces of copper, brass, or steel.

2. Repeat the procedures for lap joint brazing using silver brazing filler metal and the proper flux.

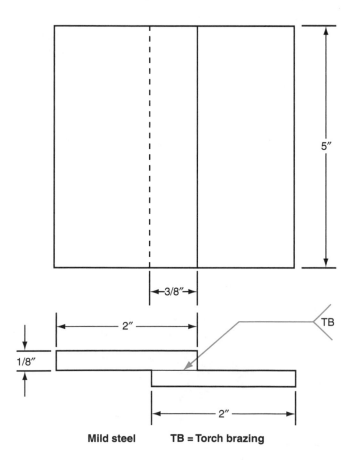

Figure 28-3. Plan for a brazed lap joint.

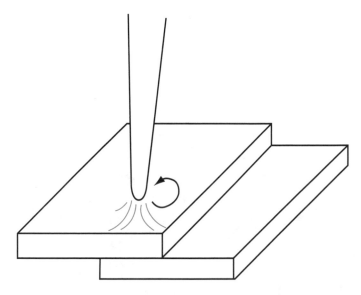

Figure 28-4. The flame is directed at the center of the lap joint. Move the flame in a circular motion to distribute the heat, taking care not to melt the base metal. When the base metal is hot enough, it will melt the filler metal.

Check Your Progress

Write your answers in the spaces provided.

1. List the six basic steps for making a brazed lap joint.

   a. _____

   b. _____

   c. _____

   d _____

   e. _____

   f. _____

2. *True or false?* The strongest brazed joint is one with a very thick layer of brazing filler metal filling a joint.

   _____

3. Where is the flux applied during brazing? _____

   _____

4. *True or false?* The torch flame is pointed at the root of the joint when brazing. _____

5. What melts the filler metal during brazing? _____

   _____

Instructor's Initials _____ Date _____

Things to Do

1. A thin line of brazing filler metal should be visible on each side of a completed braze lap joint. What are some reasons the brazing filler metal would not flow and completely fill the joint? What can be done to eliminate these problems? Discuss these questions with your instructor.

2. Perform the following experiment. Cut a 6″ (15 cm) strip of thin brass. Place it across two firebricks spaced 4″ (10 cm) apart. *Do not clean the brass surface.* Brush a silver brazing flux on the top of the strip, and heat the area from below with a neutral flame. As the flux begins to flow, add some silver brazing filler metal to the fluxed surface. Try to spread it evenly over the center of the strip. Explain the results.

Unit 29

RUNNING A BEAD
WITH BRAZING FILLER METAL

Braze Welding versus Welding

In this unit, you will gain experience running beads with brazing filler metal. Learning to form beads that are uniform in size, straightness, and appearance will take a great deal of practice. The process and materials are different from welding. In braze welding:

- The base metal is not melted.
- Brazing filler metal is used.
- Flux is used and techniques for applying the flux must be learned.
- Different torch and filler metal motions are used to form a good bead.

Running a bead with brazing filler metal is good practice for braze welding. A braze weld is produced when brazing filler metal is used to join metal pieces using a groove or fillet weld. The joint is filled with a bead similar to one made when welding. However, the base metal is not melted.

Running a bead with bronze filler metal is also good practice for hardfacing. **Hardfacing** is applying hard materials over a base metal to give the metal better wear properties.

Preparing the Equipment and Material

Before beginning the process of braze welding, the following preparations must be made:

1. Study the practice plan, **Figure 29-1.**

2. Cut or shear the material to size.

3. Obtain 1/8″ (3.2 mm) diameter bronze rod and an appropriate flux. Ask your instructor for the manufacturer's specifications for the type of alloy filler metal and flux to use.

4. Select the correct torch tip size for the base metal thickness.

Procedures for Braze Welding

Follow these procedures for running a bead with brazing filler metal:

1. Position the steel sheet flat on the welding table. Clean the length of the center area with an emery cloth or stainless steel wire brush.

2. Adjust the torch to a neutral or slightly oxidizing flame, **Figure 29-2.**

3. Heat the end of the filler metal and dip it into the flux. Apply only enough heat for a light coating of flux to stick to the filler metal.

4. Direct the flame on the right-hand edge of the steel sheet, (a left-handed person will start at the left-hand edge) and heat the steel to a dull cherry red. The width of the bead is determined by the width of the area heated to a dull cherry red. Try to keep this area and the bead approximately 1/2″ (12.7 mm) wide.

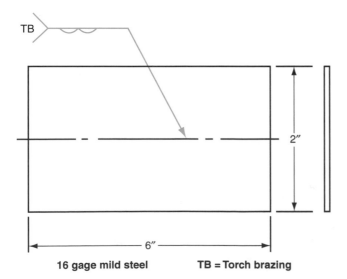

16 gage mild steel　　**TB = Torch brazing**

Figure 29-1. Plan for running a bead with braze metal.

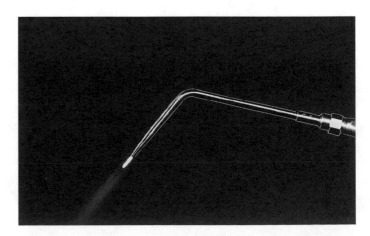

Figure 29-2. A slightly oxidizing flame. Notice the inner cone is a little shorter than the neutral flame. (Victor Equipment Co.)

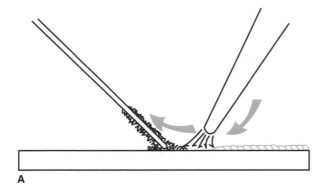

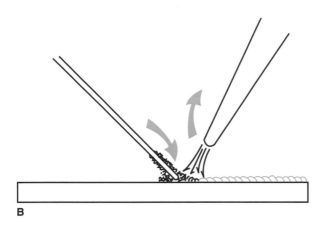

Figure 29-3. Motion of the torch and filler metal. A—The filler metal is drawn in the direction of travel, then moved up as the torch is lowered. B—The torch is pulled away as the fluxed filler metal is deposited.

5. Touch the filler metal to the heated area, and allow some flux to melt on the surface. Quickly melt off a drop of brazing filler metal, and allow it to flow to the desired bead width.

6. Add more brazing filler metal for the required bead size and begin running the bead. Direct the flame at the base metal which, in turn, melts the filler metal. Do not let the inner cone of the flame contact the molten braze metal.

7. As the braze progresses, move the torch and filler metal in an alternating up-and-down motion. See **Figure 29-3.** When removing the brazing filler metal from the pool, draw it in the direction of brazing before pulling it away from the base metal. This will cause the flux to flow ahead of the molten alloy. The flux will clean the surface and provide a path for the braze bead.

8. When the fluxed portion of the filler metal is consumed, dip the hot filler metal into the flux container to pick up additional flux. Continue brazing.

9. Examine the completed bead for straightness, uniform width and height, and an even ripple. Practice running beads until you can make them consistently.

Check Your Progress

Write your answers in the spaces provided.

1. Braze welding differs from welding in the following ways:

   a. _____

   b. _____

   c. _____

   d. _____

2. What type of motion of the torch and filler metal is used when running a brazed bead? _____

   _____

3. When brazing, adjust the torch to a neutral or slightly _____ flame.

4. What technique is used with brazing filler metal to direct the flow of the braze metal? _____

   _____

5. After running a completed bead, examine it for:

   a. _____

   b. _____

   c. _____

Instructor's Initials _____ Date _____

Things to Do

1. Practice running more beads, but deviate from proper procedures as follows:
   a. Move the torch in the standard weaving pattern, but do not use an up-and-down motion.
   b. Adjust the torch to a slightly carburizing (reducing) flame.
   c. Apply a very small amount of flux to the bronze filler metal. Do not add flux as the bead progresses.

   Carefully examine each bead after it is made. Record poor results in the space below. Discuss how proper procedures would prevent the problems from occurring.

   _____

   _____

   _____

   _____

   _____

   _____

   _____

   _____

   _____

   _____

2. Prepare a research report on the topic, "Industrial Applications of Brazing." Include types of materials and processes used, and products that are partially fabricated by brazing. Use library materials, the Internet, and manufacturers' literature for resources.

3. In the grid, sketch one of your best practice pieces. Show bead formation and how the flux and metal color appeared after brazing.

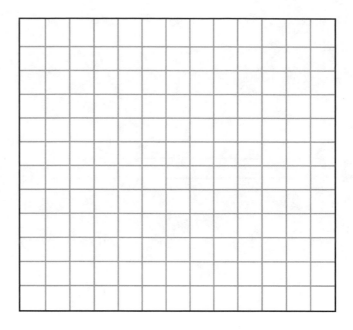

Unit 30

BRAZE WELDING LAP AND T-JOINTS

Creating Strong Joints

In this unit, you will acquire skills in braze welding lap and T-joints on steel. In braze welding, a bead is formed to create a fillet or fill a groove. Any joint design can be braze welded. The strength of a braze welded joint depends on the strength of the brazing filler metal.

Well-designed braze welded joints enlarge the surface areas, which helps increase the strength of the joint. See **Figure 30-1.** Many braze welded joints are a combination of brazing and braze welding. Greater strength is obtained with this technique.

Braze welding is different from brazing. In brazing, molten brazing filler metal is drawn by capillary action into tightly fitted joints. The strength of the joint is determined by the bond between the base metal and brazing filler metal. The thinner the joint, the greater the strength. In braze welding, a bead is formed and the strength of the joint is determined by the strength of the brazing filler metal.

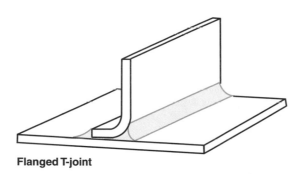

Flanged T-joint

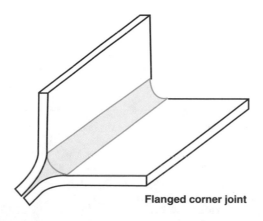

Flanged corner joint

Figure 30-1. Well-designed braze welded joints add surface area to increase joint strength.

When making braze-welded lap and T-joints, several steps should be followed to create the strongest possible joints:

1. Clean the parts.

2. Apply flux to the appropriate surfaces and edges.

3. Apply heat from the torch. Use care not to overheat the base metal.

4. Add the correct amount of filler metal to obtain complete penetration and create a properly sized fillet.

5. Clean the completed braze-welded parts.

In a lap joint, overlapping sections of the metal must be in close contact. Much of the joint strength lies in the bond between the brazing filler metal and the base metals. Some braze metal will flow into the overlapped part of the joint by capillary action, adding more strength to the joint. To create capillary action, the flame must be moved over the surface of the top piece. Additional brazing filler metal is added to create a fillet.

The technique for making a braze fillet weld on a T-joint is similar to the technique for braze welding a lap joint. However, more heat is required on a T-joint and the work angle of the torch flame is changed. Be careful not to overheat the vertical plate. Make sure penetration into the corner is complete. Maximum joint strength is achieved when the vertical plate and the bottom plate have good contact. Capillary action will draw brazing filler metal into the overlapping area and create a stronger joint.

Preparing the Equipment and Material for Lap Joints

Before beginning the process of braze welding a lap joint, the following preparations must be made:

1. Study the practice plan, **Figure 30-2.**

2. Cut the material to size, and remove all burrs and sharp edges.

3. Be sure the plates are flat and have good contact.

4. Obtain 1/8″ (3.2 mm) bronze filler rod and the proper powdered flux.

5. Select the same torch tip size used for welding a similar thickness of steel.

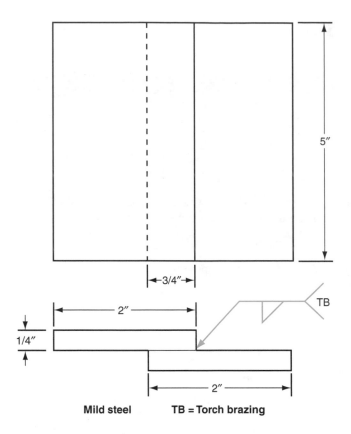

Mild steel   **TB = Torch brazing**

Figure 30-2. Plan for braze welding a lap joint.

Procedures for Braze Welding Lap Joints

Follow these procedures for braze welding a lap joint:

1. Clean the overlapping surfaces and joint edges.

2. Position the plates on the welding table so they overlap 3/8″ (4.8 mm).

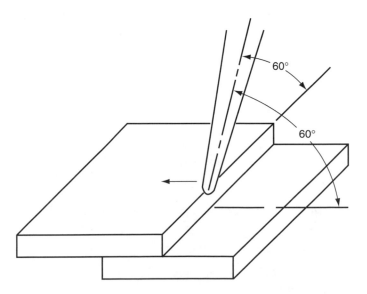

Figure. 30-3. Correct torch angle for braze welding a lap joint. Move the torch over the top surface so the braze metal is drawn into the lap joint by capillary action.

3. Adjust the torch to a neutral or slightly oxidizing flame.

4. Heat the end of the bronze rod and dip it into the flux.

5. Heat an area at the joint edge to a dull cherry red, and melt some flux from the filler metal into the corner. Heat both plates evenly.

6. Hold the torch at a 60° work angle and a 60° travel angle, **Figure 30-3.** Direct the flame into the root of the lap joint. Add brazing filler metal to create a fillet the thickness of the base metal.

7. Every few seconds, move the torch over the top piece. This will allow the braze metal to be drawn into the overlapping area by capillary action.

8. Examine the completed braze-welded lap joint.

Preparing the Equipment and Material for T-joints

Before beginning the process of braze welding a T-joint, the following preparations must be made:

1. Study the practice plan, **Figure 30-4.**

2. Cut the material to size, and remove all burrs and sharp edges.

3. Obtain 1/8″ (3.2 mm) diameter bronze rod and the proper powdered flux.

4. Select the correct torch tip size for the base metal thickness.

Procedures for Braze Welding T-joints

Follow these procedures for braze welding a T-joint:

1. Clean the plates, and position one strip vertically on the base strip to form a 90° angle.

2. Adjust the torch to a slightly oxidizing flame, and tack the plates in position.

3. Apply heat at the right-hand side of the joint. Dip the heated filler metal into the flux container, and melt some flux into the corner.

4. Hold the torch at a 45° work angle and a 75° travel angle, **Figures 30-5** and **30-6.** Begin braze welding.

5. Adjust the flame angle as necessary to deposit an equal amount of filler alloy on each leg of the joint.

6. Adjust the distance between the inner cone of the flame and the work to form a smooth bead. Complete the braze-welded joint. See **Figure 30-7.**

7. Perform the bend test used for the welded T-joint (refer to Unit 20). Inspect for penetration, fractures, and uniformity of the bond.

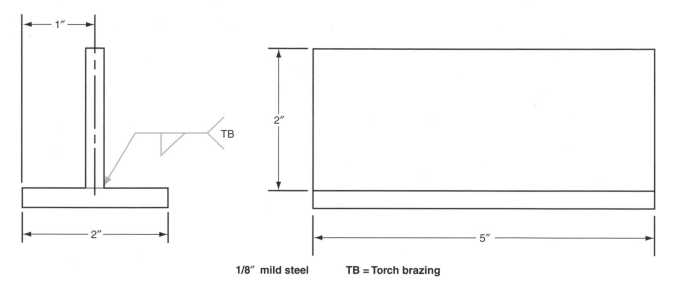

1/8″ mild steel   **TB = Torch brazing**

Figure 30-4. Plan for braze welding a T-joint.

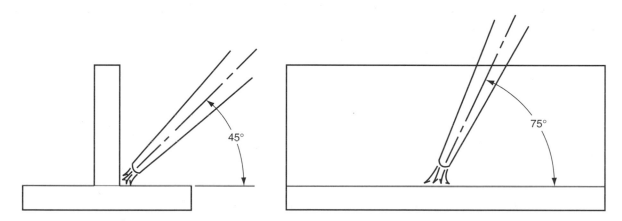

Figure 30-5. The torch flame should have a 45° work angle and a 75° travel angle.

Figure 30-6. A left-handed person braze welding a T-joint using correct torch and filler metal angles.

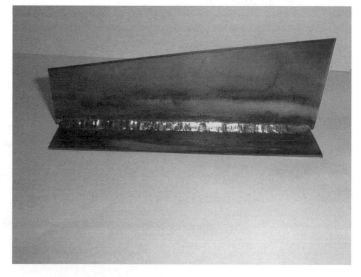

Figure 30-7. Completed braze-welded fillet joint.

Check Your Progress

Write your answers in the spaces provided.

1. A well-designed braze-welded joint increases the _____, which increases the strength of the joint.

2. In braze welding, what determines the strength of the joint?

   _____

3. List the five basic steps for ensuring the strongest possible joint:

   a. _____

   b. _____

   c. _____

   d. _____

   e. _____

4. In a lap joint, overlapping sections of the metal must be _____.

5. Why is the torch flame moved over the surface of the top piece when braze welding a lap joint? _____

   _____

   _____

6. In what two ways does the technique for making a braze fillet weld on a T-joint differ from the technique for braze welding a lap joint?

   a. _____

   b. _____

7. *True or false?* The torch tip used for braze welding is larger than the torch tip used for welding the same thickness of base metal. _____

8. The torch flame on a lap joint is directed into the corner of the joint at a work angle of *(circle letter)*:
   a. 30°.
   b. 45°.
   c. 60°.
   d. 90°.

9. The travel angle for braze welding a T-joint is *(circle letter)*:
   a. 30°.
   b. 45°.
   c. 60°.
   d. 75°.

10. The torch flame is directed into the corner of the T-joint at a work angle of *(circle letter)*:
    a. 30°.
    b. 45°.
    c. 60°.
    d. 90°.

Instructor's Initials _____ Date _____

Things to Do

1. In the grid below, sketch one of your braze-welded T-joints after the bend test. Label any weak points, such as lack of penetration or cracks. Explain what may have caused them.

   _____

   _____

   _____

   _____

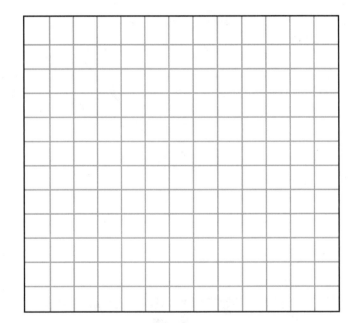

2. Continue practicing braze welding on the opposite sides of your finished pieces. Describe any problems you have braze welding the opposite side. Explain the possible causes for these problems. _____

   _____

   _____

   _____

   _____

3. Make a braze weld on a T-joint by placing the material in the position shown below. Is it easier or more difficult to braze weld in this position? Explain. _____

_____

_____

_____

_____

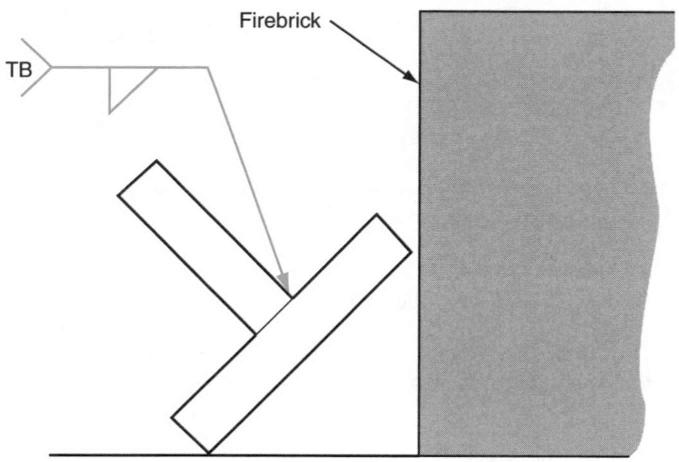

Firebrick

TB

4. Select two or three of your best braze-welded joints. Brush or spray the joints with shellac to prevent them from rusting and to bring out the details of the bead. Display the joints in your shop.

Unit 31

BRAZE WELDING BUTT JOINTS

Butt Joints and Edge Preparation

The experience you gained welding butt joints with filler metal will help you learn a similar braze-welding operation. Distinct differences exist between the processes, but many of the techniques have already been learned.

This unit will give you a chance to practice braze welding butt joints. The exercise will be done on thick steel plate to give you further experience in edge preparation.

Preparing the Equipment and Material

Before beginning the process of braze welding a butt joint, the following preparations must be made:

1. Study the practice plan, **Figure 31-1.**

2. Cut and bevel the material to size, **Figure 31-2.** Remove all burrs and sharp edges.

3. Obtain 1/8″ (3.2 mm) diameter brazing filler metal and the proper powdered flux.

4. Select the correct torch tip size for the base metal thickness.

Procedures for Braze Welding Butt Joints

Follow these procedures for braze welding a butt joint:

1. Clean the beveled edges and top faces of the plates with a wire brush or abrasive paper.

2. Position the plates flat on firebricks so the beveled edges align. Make sure the entire length of the edges are in contact.

3. Adjust the torch to a slightly oxidizing flame.

4. Heat the end of the brazing filler metal and dip it into the flux.

5. Heat the edges of the plates equally along the end of the V-groove until they turn dull cherry red.

6. Melt some flux from the filler metal into the joint before starting to braze weld. This will clean the joint and help make the bronze filler metal flow.

7. Touch the brazing filler metal onto the base metal. The heat from the base metal will melt a small amount of filler metal into the joint. Begin braze welding.

8. Keep the bead 1/2″ (12.7 mm) wide and complete the braze weld. See **Figure 31-3.**

9. Clean the braze-welded joint.

10. Test the braze weld by bending it 90°. A good braze-welded joint will show no signs of cracking or fracture.

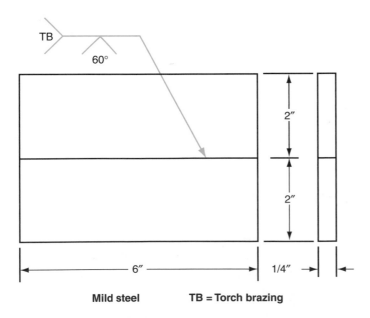

Figure 31-1. Plan for braze welding a single V-groove butt joint.

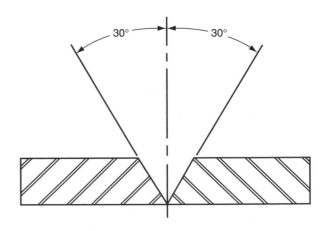

Figure 31-2. Recommended bevel angles.

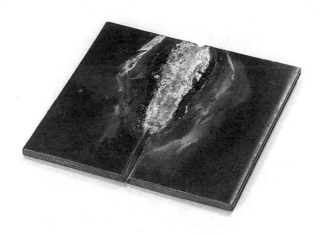

Figure 31-3. Partially completed single V-groove butt braze. Flux has not yet been removed from the braze metal.

11. Practice braze welding butt joints until they meet acceptable standards.

Check Your Progress

Write your answers in the spaces provided.

1. When braze welding a butt joint, make sure the entire length of the edges of the metal plate are _____.

2. When braze welding a butt joint, the torch flame should be adjusted to *(circle letter)*:
   a. neutral.
   b. slightly carburizing.
   c. slightly oxidizing.

3. Heat the edges of the plates equally along the end of the V-groove until the color turns _____.

4. Why should you melt some flux from the filler metal into the joint before starting to braze weld?_____
   _____

5. After a bend test, a good braze-welded butt joint will show no signs of _____.

Instructor's Initials _____ Date _____

Things to Do

1. Compare the cost of bronze brazing filler metal (RBCuZn) with mild steel welding rod. Check with a local welding supplier for up-to-date prices.

2. Obtain different types of braze-welded butt joints made by a professional welder. Compare them with braze-welded joints you have made.

3. Using steel plates of the same size, practice making square butt braze-welded joints. In the space below, explain any differences between making the square butt and the V-groove butt. _____
_____
_____
_____
_____
_____
_____

4. Sketch in the welding symbols for the braze-welded butt joints shown in the illustrations.

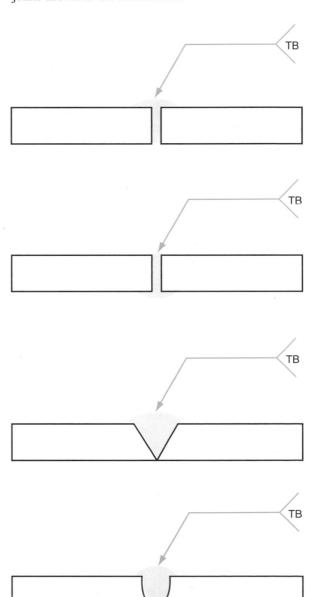

Unit 32

BRAZE WELDING CAST IRON

Purposes and Advantages of Brazing

The main purposes for braze welding cast iron are to:

- Join thick cast materials.
- Repair broken or cracked castings, **Figure 32-1.**
- Build up surfaces of worn machine parts.

Joints in cast iron can be braze welded with highly satisfactory results. If you have developed the knowledge and skills necessary to properly weld and braze weld steel, the problems associated with braze welding cast iron should not be difficult.

The challenges of working with cast iron lie in the material itself. It contains a great deal more carbon than steel. Cast iron expands and contracts considerably more than steel during heating and cooling. Therefore, the braze filler metal must expand and contract with the cast iron base metal to avoid cracks and stresses. Recommended brazing filler metals for cast iron include RBCuZn, BAg, and BNi.

With cast iron, braze welding has several advantages over welding, such as:

- Less preheating of parts.
- Faster deposition of filler metal.
- Shorter cooling period.
- Greatly reduced temperatures.

High-quality cast iron brazing flux must be used to obtain a good bond. Such fluxes usually contain an oxidizing agent to help remove excess carbon from the cast iron surface. Flux-coated filler metal should be used whenever possible to maintain a continuous flow of the molten filler alloy and flux.

Preparing the Equipment and Material

Before beginning the process of braze welding cast iron, the following preparations must be made:

1. Study the practice plan, **Figure 32-2.**

2. Cut the material to size, and bevel the edges as shown in **Figure 32-3.**

3. Obtain a proper flux for braze welding cast iron.

4. Obtain a suitable flux-coated 1/8″ (3.2 mm) diameter filler metal.

5. Select the same torch tip size used for brazing steel of the same thickness.

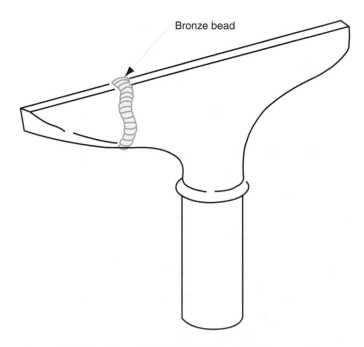

Figure 32-1. A broken cast iron lathe tool rest after braze welding. The bronze bead is ready to be machined flush with the surface.

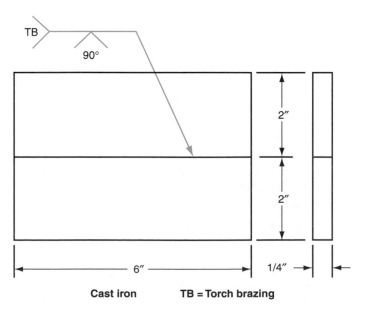

Figure 32-2. Plan for braze welding cast iron.

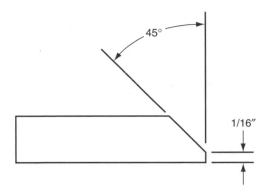

Figure 32-3. File a beveled edge on each piece. Do not grind. Grinding tends to smear the carbon on the surface and often prevents a good bond.

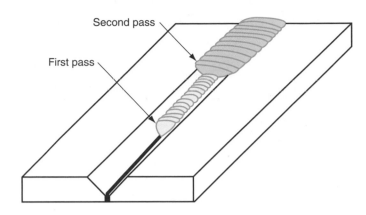

Figure 32-5. Two passes are used to braze a beveled butt on cast iron.

The first step in braze welding cast iron is tinning. *Tinning* is adding a thin layer of brazing filler metal to the surfaces to be joined. The bond created between the brazing filler metal and cast iron largely determines the strength of the joint.

Tinning is accomplished by heating the surface to a dull cherry red and laying a thin braze bead on the surface. When tinning is complete, the entire surface should be covered with filler metal. The surface of each part must be tinned. After tinning, the parts are tacked together and a braze weld is made to fill the joint.

Procedures for Braze Welding Cast Iron

Follow these procedures for brazing cast iron:

1. File a 45° angle on each piece of cast iron to be joined to create an included angle of 90°.

2. Adjust the torch to a neutral flame. Tin each of the surfaces to be joined. Hold the torch tip an inch or so from the metal and preheat the entire surface to a dull red, **Figure 32-4.** Apply a thin layer of flux-coated brazing filler metal.

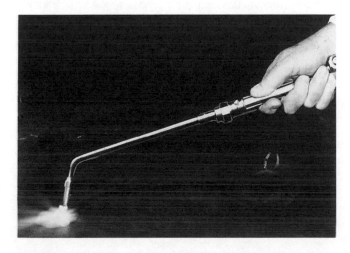

Figure 32-4. Preheating cast iron by holding the torch tip an inch or so above the surface.

3. After tinning, align the parts on firebricks so the 1/16″ (1.6 mm) flat root faces are touching. Tack both ends of the joint.

4. Start at one end of the joint and heat the parts again to a dull cherry red. Touch the flux-coated brazing filler metal to the joint. As the flux spreads, add filler metal to create a properly sized braze pool.

5. Make one pass, filling the joint approximately half the depth of the "V". Do not attempt to complete the braze in one pass. Doing so usually results in poor penetration and a weak bond.

6. After completing the first pass, deposit a second bead. This bead should evenly overlap the top edges of the beveled joint and fuse securely with the first bead, **Figure 32-5.**

Testing the Braze Weld

Test the braze weld by placing the joint in a vise with the bead positioned 1/4″ (6.4 mm) above the jaws. Hammer the upper piece on the opposite side of the bead until either the cast iron or joint breaks. If the cast iron breaks first, the brazed joint is satisfactory. If the break occurs along the bond, it usually indicates poor tinning, overheating, underheating, poor cleaning, or poor fluxing. If the bead itself breaks, it should be inspected for porous areas and impurities.

Check Your Progress

Write your answers in the spaces provided.

1. What are the main purposes of braze welding cast iron?

   a. _____

   b. _____

   c. _____

2. Cast iron contains a great deal more _____ than steel and requires the use of _____ to obtain a good bond with the base metal.

3. Name three brazing filler rods used to braze cast iron.

   a. _____

   b. _____

   c. _____

4. Braze welding cast iron has the following advantages over welding cast iron:

   a. _____

   b. _____

   c. _____

   d. _____

5. Why should flux-coated filler metal be used when braze welding cast iron? _____

_____

_____

6. To what color is cast iron heated before brazing? _____

_____

7. Define *tinning.* _____

_____

_____

8. The included angle for a beveled butt joint on thick cast iron material should be *(circle letter)*:
   a. 30°.
   b. 45°.
   c. 60°.
   d. 90°.

9. What additional step is used when braze welding cast iron that is not used when brazing or braze welding steel?

_____

10. During testing, if the cast iron breaks first, it usually indicates *(circle letter)*:
   a. poor tinning.
   b. poor fluxing.
   c. porous areas.
   d. a satisfactory joint.

Instructor's Initials _____ Date _____

Things to Do

1. Prepare a brief report on tinning. Explain the purpose of tinning and how it is done.

2. Ask your instructor for a broken or cracked cast iron part that needs repair. Properly prepare the joint, and select the correct alloy filler metal and flux. Braze the casting.

3. In the space below, sketch a beveled butt joint in 1/2″ (12.7 mm) thick cast iron that requires four passes to complete the brazing operation.

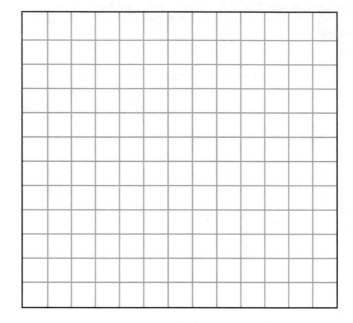

4. Report to your class on the importance of preheating and postheating thick cast iron parts to be repaired by braze welding.

5. Write to manufacturers of filler metals for brazing. Ask for specifications on alloy filler metals designed specifically for brazing cast iron. Learn about the alloy metals used in these filler metals and the purpose of each. Report your findings to the class.

6. Ask your instructor to demonstrate how a worn cast iron surface can be built up to its original shape by braze welding and machining.

Unit 33

SOLDERING

Soldered Joint Uses

In this unit, you will have an opportunity to acquire skills in making soldered lap joints on copper fittings. The same techniques can be used with other base metals. Soldering is a joining process similar to brazing. Soldering uses a filler metal with a liquidus (melting) temperature below 840°F (450°C) and below the melting point of the metals or alloys being joined. Brazing uses filler metals with a liquidus temperature above 840°F (450°C). Liquidus is the temperature at which a metal or alloy is completely liquid.

Soldering, like brazing, forms a bond between the filler metal and base metal at the surface where the two metals touch. This bond gives the joint its strength. A properly made soldered joint is airtight and watertight. Because soldering makes a leakproof joint, it is commonly used to join copper tubes that carry air and water.

Soldered joints have a clean appearance. They also have good electrical and thermal (heat) conductivity. For this reason, soldered joints are used in many electrical connections.

Soldering is usually performed on lap joints. Six basic steps must be followed when soldering:

1. Check the fit of the parts to be joined. Be sure they have good contact with each other.

2. Properly clean the metals.

3. Apply the proper flux to the surfaces to be joined.

4. Apply heat to the joint. Use care not to overheat the base metal.

5. Add an appropriate soldering filler metal. When melted, the liquid solder fills the lap joint by capillary action. *Capillary action* is when a liquid is drawn into a tightly fitted space.

6. Clean the joint after soldering.

Advantages of Soldering

Soldering has several advantages:

- Less distortion and stress are created in the parts being joined. Soldering occurs at relatively cool temperatures, and the base metals are not affected by a slight rise in temperature.

- Any of the available fuel gases produce sufficient heat to perform a soldering operation.

- In soldering, fuel gases are usually mixed with air, not pure oxygen. This eliminates the need for an oxygen cylinder, regulator, and other parts of an oxyfuel gas welding station. Also, the torch has a simpler design and is less expensive.

Filler Metals for Soldering

Soldering filler metals have melting temperatures below 840°F (450°C). They include tin, lead, antimony, silver, aluminum, indium, and cadmium. Tin/lead solders are very common and can be used to join most metal combinations. Common soldering alloys are listed in **Figure 33-1.**

Fluxes for Soldering

Cleaning the metals before joining is very important. Initial cleaning can be done using an abrasive cloth, wire brush, or chemicals. After cleaning, the appropriate flux is applied to the surfaces to be joined. The flux performs additional cleaning during the soldering process. The flux also keeps the parts from oxidizing.

Three categories of fluxes are used. Removal depends on the type of flux:

- *Organic* fluxes become noncorrosive after soldering and can be left on the parts. Nevertheless, removing them with water is recommended.

- *Inorganic* fluxes clean best but are very corrosive. These fluxes must be removed after soldering to prevent them from damaging the parts.

- *Rosin-based* fluxes are further classified as nonactive, mildly active, or fully active. Depending on the type of rosin flux used, water, alcohol, or other chemicals may be required to remove the residues.

Basic Procedures and Safety

Soldering is similar to brazing and quite different from fusion welding. Parts are cleaned, fluxed, and assembled. Heat is applied to the joint. As the temperature increases, the flux works to clean the joint. When the parts are hot enough, soldering filler metal is applied to the joint. The heat of the base metal, not the flame, melts the solder. The liquid solder is drawn into the lap joint by capillary action. Filler metal is added to fill the joint. The flame is removed and the parts are allowed to cool.

As the flux is heated, it begins to boil. Boiling can release fumes into the air. Soldering should be done in a well-ventilated

Alloy	Tin	Lead	Silver	Antimony	Cadmium	Zinc	Solidus		Liquidus	
							°F	°C	°F	°C
Tin/Lead	60	40					361	183	374	190
	50	50					361	183	421	217
	40	60					361	183	455	235
Tin/Silver	96		4				430	221	430	221
Lead/Tin/ Silver	62	36	2				354	180	372	190
	2.5	97	0.5				577	303	590	310
Tin/ Antimony	95			5			450	232	464	240
Tin/Zinc	91					9	390	199	390	199
	80					20	390	199	518	269
Silver/ Cadmium			5		95		640	338	740	393
Cadmium/ Zinc					82.5	17.5	509	265	509	265
					40	60	509	265	635	335

Figure 33-1. Common solder alloys.

area. Use a different type of solder or an air-supplied purifier or respirator if proper ventilation is not available.

Warning: Make sure excellent ventilation is available whenever solder containing cadmium is present. Cadmium is a known health hazard.

The same safety precautions observed for welding apply to all soldering operations. Protective clothing and goggles should be worn. Safe procedures for handling equipment should be observed.

Preparing the Equipment and Material

Before beginning the process of soldering, the following preparations must be made:

1. Study practice plan, **Figure 33-2.**

2. Obtain copper tubing and fittings. Cut the copper tubing to length.

3. Obtain a fuel gas air torch setup.

4. Obtain the proper soldering filler metal and flux.

Procedures for Soldering

Follow these procedures for soldering a joint:

1. Check the fit of the parts to be joined.

2. Clean the overlapping surfaces and joint edges with a wire brush or emery cloth. Clean the outside diameter of the copper tube and inside diameter of the fitting.

3. Apply flux to both surfaces. Completely cover the areas to be soldered.

4. Position the assembly so the tube is in a vertical position with the fitting on the bottom. Gravity will help the solder fill the joint.

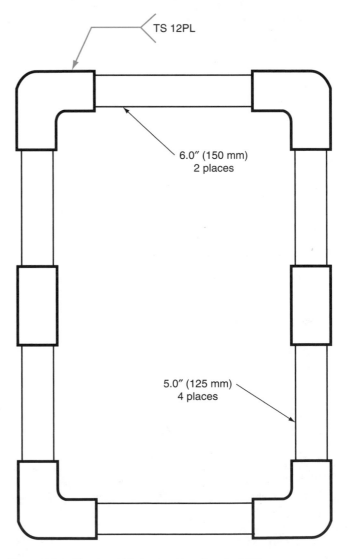

TS 12PL

6.0″ (150 mm)
2 places

5.0″ (125 mm)
4 places

Figure 33-2. Plan for soldering copper tubes and fittings.

5. Turn on the fuel gas and light the torch.

6. Use the flame to heat the center of the overlapped area. Do not direct the flame to the edge of the fitting only. Move the torch around the fitting so the entire joint area is heated. When the flux starts to boil, it is approaching soldering temperature.

7. Touch the solder to the edge of the joint. When the temperature of the metal is hot enough, the soldering filler metal will melt and be pulled into the joint by capillary action. Drag or wipe the solder around the joint.

8. Continue moving the flame and solder around the joint. The joint is filled when a thin, continuous line of solder is visible around the outside of the joint. Do not continue to add solder to attempt to create a fillet. It is not required.

9. After completing the joint, let it cool. Inspect the outside and inside (if visible). Both sides should have a thin line of solder visible.

10. Thoroughly clean the joint to remove excess flux from the surfaces.

Check Your Progress

Write your answers in the spaces provided.

1. Soldering takes place at temperatures _____ 840°F (450°C) and _____ the melting temperature of the metals or alloys being joined.

2. What gives a soldered joint its strength?_____

3. How does solder fill a lap joint? _____

4. What type of joint is used when soldering? _____

5. Which compressed gas is *not* required when soldering — fuel gas or oxygen? _____

6. List three common solders.

   a. _____

   b. _____

   c. _____

7. List the three types of fluxes used for soldering.

   a. _____

   b. _____

   c. _____

8. What category of soldering flux is the most corrosive?

9. *True or false?* Heat from the torch flame directly melts the solder. _____

10. What element present in some solders is a health hazard?

Instructor's Initials _____ Date _____

Things to Do

Contact manufacturers of soldering fluxes. Prepare a report on the types of fluxes, their performance ratings, and cleaning requirements.

Unit 34

ADDITIONAL WELDING TECHNIQUES

Common Welding Processes

The field of welding is very broad and consists of numerous welding processes. Each process has certain strengths that are useful in specific situations. However, no one welding process performs well in all situations. This unit describes the most common welding processes used in industry—arc welding and resistance welding.

Arc Welding

Arc welding consists of a group of processes that use an electrical arc to melt the base metal. An arc is created when electricity jumps across a gap between the base metal and a metal electrode. The electrode is often melted and becomes part of the weld. Not all electrodes melt, however.

In arc welding, the heat from the arc melts the base metal. Filler metal is added to the molten weld pool to fill a joint or create a fillet. The filler metal can come from an electrode melted by the arc. When a melted electrode is not part of the process, filler metal is added from a rod, similar to the way filler metal is added in oxyfuel gas welding.

Warning: The light produced in arc welding is very intense. Infrared light and ultraviolet light are capable of damaging eyes and burning skin. Eye protection must be worn when arc welding. A filter lens mounted in a welding helmet protects the welder's eyes and face. Proper protective clothing must be worn to prevent burns.

Figure 34-1. Shielded metal arc welding uses a flux-coated electrode. An arc melts the electrode and the base metal.

Shielded Metal Arc Welding

Shielded metal arc welding (SMAW) is also called stick welding. Welding takes place when an electric arc is struck (created) between an electrode and the metal to be welded. The electrode is a solid metal rod coated with a flux. SMAW electrodes are 11″ to 18″ (28 cm to 46 cm) long and 1/8″ to 3/8″ (3.2 mm to 9.5 mm) in diameter.

When the arc is struck, heat from the arc melts the base metal and the electrode. The flux around the electrode melts, adding alloying elements to the weld. As the molten weld pool solidifies, the flux solidifies on top of the weld. At this point it is called slag. The slag protects the hot weld metal during cooling.

SMAW is used for structural welding, pipe welding, repair welding, and general welding applications. See **Figure 34-1.** Many types of shielded metal electrodes are produced to weld most weldable metals. The power supply required is fairly inexpensive.

Gas Tungsten Arc Welding

Gas tungsten arc welding (GTAW) is the cleanest arc welding process and produces the highest quality arc welds. The electrode is made of tungsten and is not melted.

In GTAW, an arc is struck between the tungsten electrode and the base metal. The heat from the arc melts the base metal but not the tungsten. Additional filler metal must be added to the weld pool to create a fillet weld or fill a groove weld.

A shielding gas is used in this process. ***Shielding gas*** protects the welding area from oxygen and other atmospheric gases. The shielding gases used in GTAW are inert. This means they do not react with the welding process. Using an inert gas keeps the process very clean. The most frequently used shielding gases are argon and helium. Hydrogen can be used in combination with argon or helium when welding some base metals.

Because this welding process uses a tungsten electrode and an inert gas, it used to be called tungsten inert gas welding, or TIG welding. An even older name for this process is heliarc welding. Both names are still used, but the correct name is gas tungsten arc welding. See **Figure 34-2.**

Gas Metal Arc Welding

In gas metal arc welding (GMAW), the electrode is a continuous solid wire fed from a spool to a welding gun. The welder manipulates the gun to direct the electrode and the arc, **Figure 34-3.** The arc melts the base metal and the electrode.

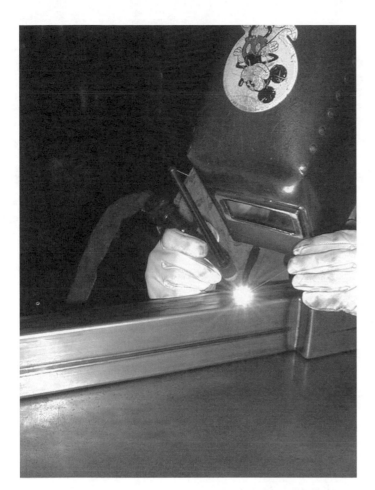

Figure 34-2. In gas tungsten arc welding, an arc is struck between a tungsten electrode and the base metal. The electrode does not melt. Filler metal is added in much the same way as in oxyfuel gas welding.

The electrode is continuously fed into the weld pool. Long welds can be made using GMAW without stops and restarts.

As in GTAW, the welding area is protected by a shielding gas. The gas may or may not be inert. Typical shielding gases include argon, helium, and carbon dioxide, used singly or in combination. Oxygen is sometimes added to the shielding gas mixture. Both carbon dioxide and oxygen allow oxygen to come in contact with the weld pool. To prevent the weld metal from oxidizing, special alloying elements must be added to the electrode to prevent contamination of the weld.

Gas metal arc welding is easy to learn. It is used in many of the applications for which shielded metal arc welding is used. GMAW will deposit more weld metal at a faster rate than SMAW or GTAW.

Flux Cored Arc Welding

Flux cored arc welding (FCAW) is very similar to gas metal arc welding. A continuous electrode is fed through the welding gun. An arc melts both the electrode and the base metal.

The biggest difference between FCAW and GMAW is the electrode. The center of the flux cored electrode is filled with a powdered flux. The flux is used to add alloying elements and deoxidizers to the weld. It also creates a protective slag over

the weld during cooling, similarly to SMAW. In addition, the flux can be used to create a protective atmosphere around the weld pool during welding.

FCAW can be done with or without shielding gas. Shielding gas is used to protect the weld area from contamination from elements in the air. In some electrodes, the flux creates a shielding gas around the weld area. Shielding gas is not required with these self-shielding electrodes. The equipment used for FCAW is similar to the equipment used for GMAW.

Resistance Welding

Resistance welding differs from oxyfuel gas welding and arc welding. A resistance weld is made on a lap joint. Two electrode tips contact the metal to be welded, one from each side of the joint. The electrodes do not become part of the weld. They are used only to locate the spot weld and carry electrical current to the weld.

Normally, thousands of amps of electrical current are passed from one electrode tip, through the metal being welded, to the other tip. The flow of electricity creates enough heat to melt the metals being welded. The surfaces of the base metals in contact with each other melt and flow together. When the

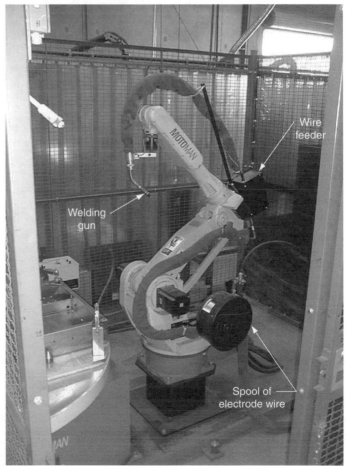

Figure 34-3. Gas metal arc welding uses a continuous electrode melted by the welding arc. The electrode is deposited into the weld pool. Here a robot holds and moves the GMAW torch.

Figure 34-4. In a resistance weld, two electrodes contact opposite sides of a lap joint. Electrical current passes between the electrodes. The metal is heated using resistance heating and a resistance spot weld is created.

liquid solidifies, a resistance spot weld is formed. The electrodes apply hundreds of pounds of force to the base metal. This force keeps the metals together and contains the liquid metal. The entire process usually takes less than one second to complete.

Resistance welding is used in automotive and appliance assembly, **Figure 34-4.** Since no arc or flame is produced, only safety glasses and gloves are worn for protection.

Other Welding Processes

Refer to the many welding processes in Figure 1-1. Some have very specific applications. Some processes, such as laser and electron beam welding, require expensive equipment. Both are high-energy-density processes; that is, a lot of energy is concentrated in a very small area. Both laser and electron beam welding processes make very narrow and deeply penetrating welds.

Welding and cutting can be done under water. A person welding under water must be a skilled welder and diver. Underwater welding and cutting are used to construct and repair oil drilling rigs, bridge columns, and pipelines.

Check Your Progress

Write your answers in the spaces provided.

1. What is arc welding?_____
   _____
   _____

2. Why must eye protection with a filter lens and protective clothing be worn during arc welding?_____
   _____
   _____

3. *True or false?* The electrode used in SMAW is melted by the arc and becomes part of the weld._____

4. A completed SMAW joint has a coating over the weld called _____.

5. *True or false?* The electrode used in GTAW is melted by the arc and becomes part of the weld._____

6. What is the purpose of an inert shielding gas? _____
   _____
   _____

7. List three shielding gases used with GMAW.

   a. _____

   b. _____

   c. _____

8. Describe how the electrode used for FCAW differs from the electrode used for GMAW. _____
   _____
   _____

9. What type of joint is used when resistance welding *(circle letter)*?
   a. square butt joint
   b. lap joint
   c. T-joint
   d. none of the above

10. Explain how a resistance weld is made. _____
    _____
    _____
    _____
    _____
    _____
    _____
    _____
    _____

Instructor's Initials_____ Date _____

Things to Do

1. Select a process from Figure 1-1. Use the Internet or library resources to obtain additional information about the process. Make a report to present to your class.

2. Visit a local welding company and observe the different welding and cutting processes used. What products were being fabricated using these processes?

Unit 35

PROFESSIONAL WELDING ADVANCEMENT

Growing Demand

Many people do not realize the extent of welding in modern industry. The term itself does not define the thousands of ways welding is used in manufacturing and construction. From the inexpensive lightbulb to computers and spacecraft, welding has applications in almost every area of the economy.

With the development of so many new welding and metal-treating processes, the demand for skilled workers continues to grow. Many career options are available in the welding industry, with preparation available through various educational programs.

This unit will explore numerous opportunities in welding. The knowledge and skill requirements for particular positions are discussed. Some job positions require minimum preparation, while others require a diploma from a high school or vocational or technical program. Some positions require a college degree to qualify for employment.

Position Descriptions

The American Welding Society (AWS) is the best source of information regarding occupations in welding. The following list of positions was compiled by AWS. Job descriptions as well as minimum education and experience requirements are given.

Welding Helper. Moves workpieces, tools and supplies; clamps workpieces to table, and into jigs or position as directed. Cleans workpieces to remove impurities such as slag, rust, and grease. Places workpieces in furnace for preheating for a specified time. Minimal education is required.

Welding Operator. Operates automatic welding equipment, including robots, set to perform a predetermined process. Ensures the machine is functioning properly, and reports any necessary adjustments or repairs to maintain the quality of the weld. Few skills are required to perform these duties. Can be trained on the job in several days. High school education is desirable but not necessary.

Welder (Manual). Includes a wide range of jobs, skills, and welding operations. An interesting position for individuals who want to work with their hands and develop skills in a variety of welding operations. Opportunities are available in all industries with specialization in such areas as structural welding, pipe welding, or aerospace welding. Opportunities also exist to do general welding in a large plant or a small repair shop. Employment can be in union and nonunion work environments.

Skilled welders are in demand. Wages can be very good, depending on skill and experience. Welders also have the opportunity for advancement in their field. With experience and further education, positions such as welding supervisor or instructor are within reach.

A welder usually needs one or two years of vocational education to secure employment. Often certification is necessary. To become certified, a welder must pass a welding test or tests that follow predetermined welding parameters. Certification proves the welder can make a desired weld. A single certification is often narrow in scope; thus, a welder may have to pass many certification tests to prove his or her skills in different areas. A good vocational education will prepare a welder to take initial certification tests.

Welder-fitter. Requires the manipulative skills of an experienced manual welder plus knowledge of blueprint reading and general welding technology. The welder-fitter must be able to follow specifications and set up work with or without jigs and fixtures.

A high school education is preferred, with emphasis on science, mathematics, and shop courses. Welding experience is required. If welder-fitters are high school graduates and pursue further education on the job, they can advance to other positions.

Specialist Welder. Many welders become experts in a specialized area, having developed a particular interest or ability during their early welding experiences. For example, a welder may develop expertise in metals that are difficult to cut or weld. Industry needs welding specialists who can solve problems. A vocational school provides good preparation. Further education can move specialist welders into supervisory positions.

Welding Supervisor. Individuals with considerable experience as welders, welder-fitters, or welding specialists are in a good position to become welding supervisors. Candidates for such jobs should demonstrate leadership qualities and the ability to responsibly supervise a team of welders. Supervisors are often called upon to teach additional skills to welders in their teams. They may be responsible for directing such operations as installation of a bridge, fabrication of steel structures, or welding processes within a company. Supervisors should be high school graduates or have received their diplomas while working in the industry.

Weld Test Inspector. Welds, tests, and inspects metal samples. Determines the suitability of welding processes. Tests metals using magnaflux or hardness tests. Inspects and tests metals for impurities and finished welds for metal properties. May read and interpret X-ray photographs of welded joints. A high school education with emphasis on science, mathematics, and shop courses is preferred. Technical or on-the-job training in various inspection techniques is required.

Welding Cost Analyst (Estimator). Provides accurate time and cost estimates for welding projects. Must be well-informed about labor, material, and time expenditures. Individuals in this position should enjoy making accurate and detailed calculations. A high school diploma with courses in science and mathematics, as well as welding experience is required.

Welding Technician. Works in the vast area between the engineer, who develops new ideas, and skilled workers, who carry out the processes. Technicians are multitalented and must enjoy problem-solving. Knowledge of drafting and design is essential. Technicians must be able to conduct laboratory tests, collect data, and write evaluation reports. They must be able to communicate with all segments of the welding industry, whether assisting an engineer or instructing an operator. Putting theory into practice is the hallmark of a technician.

Qualified technicians are in demand in the welding industry. Preparation requires a high school diploma and two years of study at a community college or technical institute.

Inspector. Maintains and judges standards of quality for operations and products. Requires training in evaluation procedures, ranging from visual inspection to more complicated nondestructive testing, such as microscopic or X-ray examination. Good interpersonal skills are essential. Inspectors are required to reject work that does not meet minimum standards without destroying confidence or creating problems for other employees.

Training is required in blueprint reading, drafting, quality control, measurement, standard testing procedures, and the use of welding symbols. A diploma from a technical or vocational high school is preferred.

First-line Supervisor. Considered part of the management team, first-line supervisors are responsible for seeing the work gets done economically and efficiently. Therefore, they must be respected and trusted by those they supervise. At the same time, they must have complete knowledge of production welding processes.

Many first-line supervisors are skilled welders who have worked their way up the ranks and improved their skills through experience and continuing education. A high school diploma with emphasis on vocational and technical courses is required.

Welding Engineer. When a new product or process is being considered, engineers are responsible for the development of an idea into a working model. They also explore the best possible methods for processing and production.

A welding research engineer combines knowledge of research techniques, the scientific method, mathematics, chemistry, and metallurgy to investigate new materials and processes. The research engineer must be able to recognize new discoveries and evaluate them for possible use in industry. A working knowledge of design principles, electronics, mechanics, metallurgy, and construction is necessary. A thorough knowledge of welding processes is required as well. Research often opens the door to planning new procedures, products, and processes.

Most engineers are college graduates. Some hold master degrees or doctorates in specialized fields. The welding industry especially needs engineers in metallurgy, research, and development.

Sales Engineer. Must have a good understanding of the welding equipment being sold. May be called upon to demonstrate products, assist customers, and attend sales meetings. Must stay keenly aware of the market trends. Sales engineers must understand all phases of the welding industry. They must be able to answer questions about products and processes, describe equipment capabilities, and help customers solve their production needs. A college education is highly recommended. Sales experience and knowledge of the welding industry is very desirable. Good verbal and written communication skills are required.

Educational Opportunities

As new technologies are applied, the field of welding continues to expand. Staying current with changes in the industry can only be accomplished through education and retraining.

Welding techniques are learned through experience and teachers. School is one of the best places to initially learn the skills necessary to be a welder. Education will continue throughout your working career in the form of on-the-job training, seminars, classes, reading, company visits, and formal or informal discussions with others in the field.

The American Welding Society publishes a monthly magazine and has local chapter meetings. Numerous books and journals on welding are published by other sources as well. Visits to local welding distributors to learn about new products or to local businesses that perform welding can also expand your knowledge. Formal and informal education is available through technical societies and professional organizations. Training is often in the form of seminars offered for a few hours to a few days. A specific topic is usually presented. Local chapters of AWS meet in most cities in the United States, giving you the opportunity to network with colleagues in the welding field.

Many community colleges offer programs in welding. Community colleges enable you to learn new welding processes and become certified, a distinction that can lead to advancement. Colleges and universities offer classes and degree programs in welding, welding technology, and welding engineering. As you can see, many opportunities for learning are available throughout your career.

Check Your Progress

Write your answers in the spaces provided.

1. A person who operates automatic welding equipment set to perform a predetermined process is known as a(n) _____.

2. How does a manual welder become certified? _____
_____

3. List three job requirements of the welder-fitter.

   a. _____

   b. _____

   c. _____

4. Give an example of an area of expertise developed by a specialist welder. _____

   _____

5. What position sees that work gets done economically and efficiently, has complete knowledge of welding processes, and is considered part of the management team? _____

   _____

6. What position in the welding field evaluates completed welds by visual inspection and nondestructive evaluation?

   _____

7. What is the major responsibility of the welding cost analyst? _____

   _____

8. What welding occupation is described by the following: Knowledge of drafting and design is essential. Must be able to conduct laboratory tests, collect data, and write evaluation reports. Puts theory into practice. *(Circle letter.)*
   a. welding technician
   b. first-line supervisor
   c. specialist welder
   d. inspector

9. In what three areas are welding engineers especially needed?

   a. _____

   b. _____

   c. _____

10. List three ways to continue your education in the welding field.

    a. _____

    b. _____

    c. _____

Instructor's Initials _____ Date _____

Things to Do

1. Use the Internet to obtain online information about opportunities in the welding field. Start by choosing a search engine and keying in the words American Welding Society.

2. Write to large manufacturers of welding equipment and supplies. Many have their own training programs. Ask for literature describing training and a list of welding occupations for which they offer preparation.

3. Visit a local job site or custom welding shop. Ask the manager what qualifications and responsibilities are required for employment as a welder. Prepare a report to present to your class.

4. Invite an experienced welder to speak to your class about his or her job.

5. Survey your community to learn the types of welding being done in various shops and companies. Make a list of all the welding processes you find being performed.

6. Find out as much as you can about welding occupations using the *Occupational Outlook Handbook* in the library, as well as information from labor unions or the guidance office at your school.

Conversion Table
US Conventional to Metric

When you know:	Multiply by: * = Exact		To find:
	Very accurate	Approximate	
Length			
inches	* 25.4		millimeters
inches	* 2.54		centimeters
feet	* 0.3048		meters
feet	* 30.48		centimeters
yards	* 0.9144	0.9	meters
miles	* 1.609344	1.6	kilometers
Weight			
ounces	*28.349523125	28.0	grams
ounces	* 0.028349523125	0.028	kilograms
pounds	* 0.45359237	0.45	kilograms
short ton	* 0.90718474	0.9	tonnes
Volume			
teaspoon		5.0	milliliters
tablespoon		15.0	milliliters
fluid ounces	29.57353	30.0	milliliters
cups		0.24	liters
pints	* 0.473176473	0.47	liters
quarts	* 0.946352946	0.95	liters
gallons	* 3.785411784	3.8	liters
cubic inches	* 0.016387064	0.02	liters
cubic feet	* 0.028316846592	0.03	cubic meters
cubic yards	* 0.764554857984	0.76	cubic meters
Area			
square inches	* 6.4516	6.5	square centimeter
square feet	* 0.09290304	0.09	square meters
square yards	* 0.83612736	0.8	square meters
square miles		2.6	square kilometer
acres	* 0.40468564224	0.4	hectares
Temperature			
Fahrenheit	* 5/9 (after subtracting 32)		Celsius

Conversion Table
Metric to US Conventional

When you know:	Multiply by: * = Exact		To find:
	Very accurate	Approximate	
Length			
millimeters	0.0393701	0.04	inches
centimeters	0.3937008	0.4	inches
meters	3.280840	3.3	feet
meters	1.093613	1.1	yards
kilometers	0.621371	0.6	miles
Weight			
grams	0.03527396	0.035	ounces
kilograms	2.204623	2.2	pounds
tonnes	1.1023113	1.1	short ton
Volume			
milliliters		0.2	teaspoons
milliliters	0.06667	0.067	tablespoon
milliliters	0.03381402	0.03	fluid ounces
liters	61.02374	61.024	cubic inches
liters	2.113376	2.1	pints
liters	1.056688	1.06	quarts
liters	0.26417205	0.26	gallons
liters	0.03531467	0.035	cubic feet
cubic meters	61023.74	61023.7	cubic inches
cubic meters	35.31467	35.0	cubic feet
cubic meters	1.3079506	1.3	cubic yards
cubic meters	264.17205	264.0	gallons
Area			
square centimeters	0.1550003	0.16	square inches
square centimeters	0.00107639	0.001	square feet
square meters	10.76391	10.8	square feet
square meters	1.195990	1.2	square yards
square kilometers		0.4	square miles
hectares	2.471054	2.5	acres
Temperature			
Celsius	*9/5 (then add 32)		Fahrenheit

Glossary of Technical Terms

A

accurate: Produced within specified tolerances.

acetone: Flammable liquid that serves as a dissolving agent in acetylene cylinders.

acetylene: Combustible gas made up of carbon and hydrogen, and produced by combining calcium carbide and water.

align: Bring into adjustment to established points.

alloy: Two or more metals melted together to form a new metal.

alloying element: Substance added to a single metal to change the properties of the metal.

annealing: Heating metal to a specified temperature, then allowing it to cool slowly to remove stresses and produce softness.

arc welding: Process that uses an electrical arc between an electrode and a base metal to melt the base metal.

arrow side: Information in a welding symbol that appears below the reference line and relates to the side of the weld joint to which the arrow points.

axis of weld: Centerline through the length of a weld.

B

backfire: Popping sound that occurs when the torch flame goes out. The flame may or may not relight.

backhand welding: Technique in which the torch flame is directed opposite the direction of travel over the completed weld.

backing: Any material used to support or back up a joint during welding.

base metal: Metal or metal parts to be welded or cut.

bead: Contour or deposit of metal left by any of the welding processes.

bend test: Destructive test that bends a welded part to evaluate the quality of the weld.

bevel: Angle prepared on the edge of a part to be joined by a welding process.

braze welding: Joining process that uses filler metal with a melting temperature above 840°F (450°C) but below the melting temperature of the base metal. The filler metal is used to fill a groove joint or form a fillet. The filler metal is not distributed by capillary action.

brazing: Joining process that uses a filler metal with a melting temperature above 840°F (450°C) but below the melting temperature of the base metal. The filler metal is drawn into a joint between closely fitting parts by capillary action.

brazing filler metal: Filler metal used in brazing operations.

brittleness: A characteristic of some metals that causes the metals to be easily broken under various conditions.

burr: Sharp edge left on metal after cutting, shearing, or machining operations. Burrs should be removed to avoid injury.

butt joint: Joint formed where edges of metal to be joined butt up to one another.

C

capillary action: Property of a liquid that allows it to be drawn (pulled) into tightly fitting spaces.

carbon steel: *See* low-, medium-, and high-carbon steel.

carburizing flame: Oxyfuel gas flame in which some of the fuel gas is not burned in the flame; excess carbon still remains.

clearance: Distance one object is spaced from another. Generally used in welding to allow for penetration.

color test: Identifies nonferrous metals such as brass, bronze, and aluminum by their colors.

combustion: The process of burning.

complete fusion: Condition in which base metal and filler metal have melted and fused over the entire area exposed for welding.

concave weld: Weld face that curves inward.

concentric: Having a common center or axis.

cone: Pointed part of a gas flame close to the orifice of the torch tip.

continuous weld: Weld that extends the entire length of an object without interruption.

contour: Irregular shape or outline of an object.

convex weld: Weld face that curves outward.

corner joint: Weld joint formed where the edges of two pieces are aligned to form a corner.

cover glass: Clear glass fitted over the filter lens in a pair of goggles or helmet to protect the welder from spatter.

crater: A depressed area at the end of a weld that is not properly filled with filler metal.

cutting attachment: Device fastened to a welding torch to convert it to a cutting torch.

cutting oxygen: High-pressure oxygen from a cutting tip that burns or cuts a kerf in the base metal.

cutting tip: Torch tip specially designed for preheating and cutting metal.

cutting torch: Torch designed specifically for flame cutting metal.

cylinder: Container that holds a compressed gas, such as oxygen or acetylene.

cylinder pressure gauge: A device attached to a regulator that shows pressure inside a cylinder.

D

defect: A flaw that is larger than allowed by a governing specification. A defect renders the weldment unusable unless repaired.

density test: Identifies a metal by determining its relative weight.

deposited metal: Metal added during a welding process.

depth of fusion: Distance a weld penetrates into the base metal from its original surface.

destructive evaluation: Testing of a weld that results in damaged parts that are not usable.

distillation: Boiling and separating a single component out of a liquid mixture.

ductility: Capability of a metal to be bent, drawn, or hammered into a new form.

E

edge joint: Joint formed where the surfaces of the pieces to be joined are in contact and edges are aligned.

edge preparation: Contour prepared on the edge of a part for a specific welding job.

expansion: Increase in any dimension of a metal part, usually caused by increased temperature.

F

face of weld: Exposed surface of a weld on the side where welding was performed.

ferrous metals: Metals or alloys whose major ingredient is iron.

filler metal: Metal added when welding, brazing, braze welding, or soldering. *See* welding rod.

fillet test: Destructive test that bends one piece of a joint downward onto the other piece to expose penetration in order to evaluate the quality of the weld.

fillet weld: Type of weld used to join two pieces that form an angle of approximately 90°.

filter lens: Used in goggles and helmets to protect the welder's eyes from harmful ultraviolet and infrared light rays.

fixture: Device for holding work in position during welding.

flame cutting: High-temperature oxidation process used to cut metal. Heat produced by the flame is followed by a stream of oxygen from the torch. A narrow slot in the base metal is burned away to form the cut.

flammable: Something that can be burned.

flashback: Dangerous condition in which burning gases travel back from the torch into the hoses, regulators, or even cylinders.

flashback arrestor: Device used to stop burning gases from moving from the torch into the hoses.

flaw: An imperfection in the weld or base metal that does not match the overall structure of the base metal. A flaw is also called a discontinuity.

flux: Chemical compound used to clean base metal and allow braze or solder to flow onto the base metal.

forehand welding: Technique in which the flame is pointed in the direction of travel.

fusing: Heating two or more metals or nonmetals until they become liquid, then allowing them to join and solidify.

fusion welding: Process that uses heat, often from a flame or arc, to melt a portion of each of the materials to be joined. The liquid portions flow together, then cool and solidify to form a weld.

fusion zone: Cross-sectional area of melted base metal.

G

gas pocket: Cavity in a weld caused by gases trapped during welding.

groove weld: Used to join the edge(s) of metal in a butt joint. Filler metal is required to fill the groove.

H

hardening: Heating and quenching of specific iron alloys to produce a hardness greater than the untreated metal.

hardness test: Identifies grades of steel based on how easy or difficult it is to cut the steel with a file.

heat-affected zone: Portion of the base metal that has not been melted but whose structural properties have been changed by the heat of welding or cutting.

heat treatment: Carefully controlled heating and cooling of a metal or alloy in a solid state to bring about such desirable conditions as hardness, softness, or toughness.

high-carbon steel: Steel containing 0.55% to 0.80% carbon.

horizontal position: Welding with the axis of the weld in the horizontal plane and the base metal in or near the vertical plane.

hot shortness: A characteristic of some nonferrous metals whereby the metal looses its strength at elevated temperatures.

I

ignition temperature: Temperature to which a base metal must be heated before burning can begin. Above ignition temperature, high-pressure oxygen will burn the heated base metal.

included joint angle: Total angle of a groove weld to be filled.

incomplete fusion: Weld in which areas of the base metal are not melted or have not fused with the weld metal.

injector-type torch: Torch used for oxyacetylene welding in which the pressure of the acetylene is very low and drawn into the torch by the oxygen.

inspection: Measuring and checking finished parts to determine if flaws and defects exist and if specifications have been met.

intermittent weld: Pattern of a weld in which the continuity of the run is broken by unwelded portions.

in-tip mixers: Assembled torch tips, each with its own mixing chamber.

J

joint design: Proper shape of the surfaces where parts are to be joined by welding or brazing. Required dimensions are usually specified.

K

kerf: Space or width of the cut made by the oxygen stream during flame cutting.

keyhole: Technique used to obtain complete penetration in a butt joint weld. The heat source melts the base metals back slightly, and the molten weld metal fills in behind the opening. The opening looks like a keyhole.

kilopascal: Metric measure of pressure expressed in thousands of pascals. Abbreviated kPa.

L

lap joint: Joint formed where the metal surfaces overlap each other.

lay out: To locate and scribe points for welding or cutting operations.

leg of fillet weld: Distance from the toe of a fillet weld to the root of the joint.

low-carbon steel: Steel containing less than 0.30% carbon.

M

magnetic test: Method of separating ferrous and nonferrous alloys. Metals containing iron will be attracted by a magnet; nonferrous metals will not.

malleability: A property of metal determined by the ease with which it can be shaped by mechanical working.

manual welding: Welding done entirely by hand.

medium-carbon steel: Steel containing 0.30% to 0.55% carbon.

melting rate: Weight or length of filler metal melted in a given period of time.

melt-thru: Complete joint penetration of filler metal in a joint welded from one side, providing visible root reinforcement.

mixing chamber: Section of a welding or cutting torch in which fuel and oxygen gases are mixed for combustion.

molten: Liquid portion of a material that has been heated until it melts.

N

neutral flame: Oxyacetylene or oxyfuel gas flame in which all the fuel gas and oxygen are burned and joined in the flame.

nondestructive evaluation: Inspection or testing of a weld that does not result in damaged parts.

nonferrous metals: Metals or alloys in which iron is not the major ingredient. Iron may be present as an alloying element.

O

orifice: Small opening in the end of a torch tip through which gases flow.

other side (of the weld joint): Information in a welding symbol that appears above the reference line. This information relates to the side of the weld joint opposite the arrow.

out-of-position welding: Welding in a position other than the flat position. *See* horizontal position, vertical position, and overhead position.

overhead position: Welding on the underside of a weld joint.

overlap: Extension of filler metal beyond the bond at the toe of a weld.

oxidation: Process in which oxygen combines with other elements to form oxides.

oxidizing flame: Oxyacetylene or oxyfuel gas flame in which all the oxygen is not burned in the flame. Excess oxygen remains.

oxyacetylene torch test: Performed on ferrous metals to determine if they are weldable. A small molten pool is made in the metal. If the steel is weldable, the pool will emit few sparks and be fluid instead of sluggish.

oxyacetylene welding: Process that burns acetylene and oxygen in a flame to create a source of heat. Welding rod may or may not be used.

oxyfuel gas welding: Welding process that burns a fuel gas and oxygen in a flame. The flame provides the heat for welding.

oxygen: Colorless, tasteless, odorless gaseous element. Air contains approximately 22% oxygen. In flame welding, oxygen is combined with a fuel gas, the combination is burned and heat is generated.

P

parent metal: *See* base metal.

pass: A single welding operation along a joint or weld deposit that results in a weld bead.

penetration: Distance the fusion zone extends below the surface of welded parts.

porosity: Voids or gas pockets trapped in metal.

positive pressure–type torch: Torch used for oxyacetylene welding in which the pressures of both the oxygen and acetylene gases are the same. Also called an equal pressure torch.

post-heating: Heat applied to metal after welding or cutting to slow cooling and minimize cracking or stresses.

preform: Brazing filler material or solder produced in a given form or shape for a specific application.

preheating: Heat applied to metal prior to welding, brazing, or cutting.

pressure regulator: Device that reduces high-pressure gas to working pressures.

psig: Abbreviation for pounds per square inch gauge, a unit of measurement for pressure in the US conventional system.

Q

quenching: Rapid cooling of heated metals by bringing them in contact with liquids or gases.

R

reference line: Horizontal line on a welding symbol.

regulator wrench: Tool used for tightening a pressure regulator after it has been hand-threaded onto a cylinder nozzle.

ripple: Pattern on the surface of a completed weld.

root opening: Spacing at root of joint between parts to be joined.

S

slag: Melted material that hangs on the bottom of a flame-cut kerf. Also the glass-like material that covers some types of completed arc welds.

slag inclusion: Trapping of nonmetallic solid materials within a weld or between the weld metal and base metal.

soapstone: Soft stone with a soapy texture composed essentially of talc and mica-related minerals.

soldering: Joining process in which the metal pieces to be joined are heated to a temperature below 840°F (450°C) and below the melting point of the base metal. The molten solder is drawn into the joint by capillary action.

spark lighter: Device used to ignite the gases from a welding or cutting torch.

spark test: Identifies different grades of steel and cast iron based on the color, length, and shape of the stream of sparks produced when a piece of steel or cast iron is pressed against a rotating grinding wheel.

spatter: Small molten metal particles that leave the welding area and do not become part of the weld.

standard: Document or accepted practice used to establish guidelines for design, quality, workmanship, measurement, and other areas in the welding field.

straightedge: Precision instrument for checking the accuracy of flat surfaces.

stress: Intensity of internal forces at a given point in a metal.

stringer bead: Particular type of weld bead made without a weaving motion.

T

tack weld: Small weld used to hold pieces together before welding the joint.

template: Device used as a pattern or guide.

tensile strength: Maximum load a piece can maintain under tension without breaking or failing.

tension: Force or load that pulls on a material.

throttle: Making the final flow adjustment to gases being fed to the torch to produce the proper flame.

tip cleaner: Tool used to remove slag or foreign material from the orifices of welding or cutting tips.

T-joint: Weld joint formed where the edge of one piece is placed on the surface of another piece forming the shape of a T.

toe of weld: Intersection between the face of a weld and the base metal.

tolerance: Allowable deviation from a basic dimension.

travel angle: The angle of the torch tip to the workpiece surface measured from the surface, parallel to the direction of travel, to the torch.

U

undercut: Groove melted into the base metal next to the toe of the weld which is not filled by weld metal.

V

vertical position: Welding with the axis of the weld and the base metal in or near the vertical plane.

visual inspection: Viewing a weld and evaluating its quality.

W

weaving: Technique of metal deposition in which the torch is moved in an oscillating pattern.

weld bead: Area that is melted and solidified during the welding process.

weld joint: Area where the ends, surfaces, or edges of two pieces of metal are to be joined with a weld.

weld metal: Any metal melted during welding.

weld pool: Molten area created during a welding process.

weld root: Points opposite the weld side where the weld intersects base metal surfaces.

weld symbol: Part of the welding symbol indicating the type of weld to make.

welder: A person skilled in performing manual or semiautomatic welding operations.

welding: Process of joining pieces, usually by heating parts to be joined until they melt and the liquid portions flow together.

welding machine: Equipment used to perform various welding operations. Specific settings and adjustments are made before welding.

welding rod: Solid metal added to a weld to fill a groove weld or form a fillet weld. Welding rod is often called filler metal.

welding symbol: Symbol on a drawing indicating the location and type of weld to make.

welding tip: Part of the torch assembly where mixed gases are emitted and burned, producing the high-temperature flame.

weldment: Assembly of parts joined by a welding process.

work angle: Angle of the torch tip to the workpiece surface measured from the surface, perpendicular to the direction of travel, to the torch.

working drawing: Provides the information necessary for a person to make and assemble a mechanism or product.

working pressure gauge: Device attached to a regulator that shows outlet pressure from the regulator. This is the pressure of the gas being delivered to the torch tip.

X

x-ray test: Nondestructive inspection process used to detect internal flaws in metal parts.

Index